Methods and Tools for Computer-aided Design of Multidimensional Filters in Signal Processing

IEEE Press
445 Hoes Lane
Piscataway, NJ 08854

Methods and Tools for Computer-aided Design of Multidimensional Filters in Signal Processing

Andrzej Handkiewicz
Jacob of Paradise Academy
Gorzow Wlkp., Poland

IEEEPress
WILEY

Published by John Wiley & Sons, Inc., Hoboken, New Jersey.
Published simultaneously in Canada.

Library of Congress Cataloging-in-Publication Data applied for:
Hardback ISBN: 9781394263622

Cover Design: Wiley
Cover Image: © YAY Media AS/Alamy Stock Photo

Set in 9.5/12.5pt STIXTwoText by Straive, Chennai, India

To my grandchildren:
Jadwiga, Jan, Tymoteusz, Weronika,
Tytus, Karol, Leon, Jędrzej,
"…don't be afraid, sail out into the deep…" (JPII).

Contents

About the Author

Andrzej Handkiewicz is a professor at the Jacob of Paradise Academy in Gorzow Wlkp., Poland. After one year of research, he left the Swiss Federal Institute of Technology in Zurich in the early 1990s and returned to the Poznań University of Technology to become vice dean of the Faculty of Electrical Engineering, a position he held in the years 1993–1999. Since 2005, he has been the vice dean of the Faculty of Computing for eight years. He is a member of many Polish and international scientific societies (two sections of the Electronics and Telecommunications Committee of the Polish Academy of Sciences and three IEEE societies). He is the author of *Mixed-Signal Systems: A Guide to CMOS Circuits Design*, published by Wiley-IEEE in 2002. He has also published over 100 technical papers in the areas of mixed-signal circuits, switched capacitor and switched current filters, and digital filters for image compression and processing.

Foreword

There are many excellent monographs devoted to signal processing, in which this issue is discussed at various levels of advancement. These include both textbooks that have been used by entire generations of students and items based on the most complex mathematical apparatus. In terms of the above-mentioned features, the new book proposed to readers does not try to compete with the existing rich literature. Therefore, I tried not to duplicate content that already appears in the existing literature. I focused on three aspects that, in my opinion, may be of interest to the reader. The first one is the use of gyrator-capacitor prototype circuits (gC) to design filters. This design trend has existed since the dawn of filter theory but is based on reactance networks, in particular those with a ladder structure. The monograph shows that the use of gC circuits as prototype circuits creates new perspectives in the design of multiport filters and especially multidimensional ones. The second trend in the monograph concerns computer-aided design tools, which were developed with my participation and are open access, and their description in the monograph not only facilitates their use but also enables their creative development. The third aspect, rather rarely discussed in books related to signal processing, concerns the hardware implementation of the designed filters. This is not only about digital implementation but also about the use of techniques such as OTA-Cs and switched-currents (SI). In my opinion, these are prospective techniques related to the development of field effect transistors (FET) in 2 nm technology.

Preface

It would be difficult to point to an area in which signal processing does not play a significant role. Because it involves the processing and transmission of information, it is important not only in technical disciplines but also generally with all activities demonstrated by people organized in specific social groups. An image, or a video signal, plays a special role in human perception. Cave paintings from several dozen thousand years ago indicate the huge role of this type of signals in the way of understanding and explaining the reality surrounding humans. The influence of pictorial perception of reality on our consciousness is especially visible nowadays thanks to smartphones and other handheld devices that we currently have. Algorithms and techniques for processing video signals play a special role in the development of these devices.

The development of video compression standards is closely related to hardware feasibility. The continuous growth of memory sizes and processor speeds and other hardware such as field programmable gate arrays (FPGAs) enables the implementation of algorithms of increasing computational complexity. Open access to compression standards plays an important role in the development of compression methods. An example is the AV1 standard developed by the Alliance for Open Media consortium. The AV1 codec is characterized by a balance between compression performance and the current level of development of integrated circuit technology. The technological aspect is strongly emphasized in the presented monograph. The digital sphere dominates signal processing. Digital circuit designers have a wide selection of computer-aided design tools here. The bottleneck is the design of analog circuits, the use of which is necessary at the border between the world of discrete and continuous time signals. The book discusses two techniques for implementing prototype gC circuits and associated design tools. These are OTA-C and switched-current (SI) techniques, which seem to be particularly useful from the point of view of currently developed field effect transistors (FET) in the nanometer scale.

July, 2014
Gorzow Wlkp., Poland

Acknowledgments

There are many people to whom I owe gratitude and thanks. These include all those who introduced me to the secrets of circuit theory and signal processing. It is impossible to list them all. I will stop at three professors who are also known to a wider audience. The first of them is Prof. Stanisław Bellert, with whom I was able to cooperate for many years as a doctoral student in Warsaw. He was a role model for me not only because of his extensive knowledge but also as a man with great personal culture. The next one is Prof. George Moschytz, with whom I was able to expand my knowledge and skills while staying in Zurich for several months. Being aware that this list is very limited, I would also like to mention Prof. Alfred Fettweis, whom I could visit in Bochum and whose scientific works were an inspiration for my scientific activity.

I also owe a lot to my family, who have supported me throughout my professional career, in performing both everyday teaching and scientific duties. I am especially grateful to my wife, Hanna, for the patience and understanding she showed me.

I am also grateful to many of my younger colleagues. Cooperation with them has always been an opportunity to expand my scientific horizons. Special thanks are due to those with whom I cooperated as a supervisor of their doctorates. The effects of this cooperation are largely presented in this monograph. These are Paweł Śniatała, Marcin Łukowiak, Radosław Rudnicki, Marek Kropidłowski, Piotr Katarzyński, Szymon Szczęsny, and Mariusz Naumowicz. Two of them are professors at the Poznań University of Technology, and one is a professor at the Rochester Institute of Technology. The others also have significant achievements not only in scientific activities but also in management and industry. Their achievements make me proud.

I would also like to thank the reviewers whose comments brought the content of the monograph closer to the readers' expectations. My sincere thanks go to the entire Wiley team, with whom I had the opportunity to improve editorially the book.

Andrzej Handkiewicz

Acronyms

ADC	analog-to-digital converter
ADST	asymmetrical discrete sine transform
BIBO	bounded input/bounded output stability
CKL	current Kirchhoff's law
CMRR	common mode rejection ratio
DCT	discrete Cosine transform
$\mathcal{F}$	Fourier transform
$\mathcal{L}$	Laplace transform
FET	field effect transistor
FI	frequency-interleaved converter
FIR	finite impulse response
FPGA	field programmable gate array
gC	gyrator-capacitor circuit
I_D	drain current
IIR	infinite impulse response
KLT	Karhunen–Loeve transformation
LO	local oscillator
LUT	lock up table
MDp	class of MultiDimensional polynomials
MV	motion vector
NS	nano-sheet
OTA-C	operational transconductance amplifier, capacitor circuit
RAM	random access memory
RF	radio frequency
RGB	red, green, blue color space
RLC	resistance, inductance, capacitance circuit
$S_{x_i}^{\lvert H^a \rvert}$	sensitivity of transfer function magnitude
S_c	scattering matrix
SNc	structural number class

SC	switched-capacitor
SI	switched-current
V_{DS}	drain-source voltage
V_{GS}	gate-source voltage
VHDLA	very (high speed) hardware description language – analog (and mixed-signal)
V_T	threshold voltage
$\mathcal{Z}$	Z transform

Introduction

Signal processing is mostly performed using digital circuits. The development of digital technology, resulting from technological progress in the implementation of field-effect transistors (FET), allows the production of efficient processors and huge memories. Thanks to this, it is possible to implement very complex signal processing algorithms, in particular video sequences, in real time. However, the signals we perceive and generate in the environment immediately surrounding us are analog. This makes analog circuits unavoidable in signal processing systems. Even though these circuits make a small contribution to the overall system, their design is the most time-consuming process.

The monograph is devoted to the design of circuits for signal processing in the entire spectrum of their applications, starting from single filters through filter banks, ending with image and video filters. The presented design strategy is based on gyrator-capacitor (gC) circuits as prototype networks for these filters. Thanks to this approach, it is possible to implement the designed systems both as analog filters working in continuous time with transconductance amplifiers and capacitors (OTA-C) and made using switched-capacitor (SC), switched-current (SI), and digital techniques. Particular attention was paid to the last two techniques, SI and digital, suitable for manufacturing as integrated circuits in the most advanced nanometer-scale technologies.

Technological development is focused on digital systems, therefore full automation of their design is possible. This direction of action is illustrated in the monograph by implementing a video filter on an FPGA. It was assumed that the designer has a wide range of tools oriented to digital circuits, and in the book the emphasis is on analog circuit design. Examples of filters implemented using SI and OTA-C techniques are included. These techniques seem to be the most suitable for fabrication using the latest technologies. The mathematical and algorithmic foundations of appropriate design methods and tools are presented. However, detailed familiarization with the presented algorithms is possible by providing the reader with both compiled and source versions of individual programs. The monograph contains five chapters.

The first chapter presents the "classic" approach to filter design. Filters are described in the complex domain *s*. The text description of the gC circuit in VHDLA is illustrated with an example of a simple third-order elliptic filter. The operation of the **gCstudio** environment for symbolic analysis of a prototype circuit is shown. Options of the **optimizer** tool, another fundamental **NANO-studio** environment, enabling the determination of the g and C parameters of the prototype circuit are discussed. Sections 1.2 and 1.3 of this chapter show the capabilities of the new design tools using advanced filter examples: the seventh-order elliptic and the so-called filter pair. Section 1.4 is devoted to the problem of low sensitivity of prototype filters to changes in their parameters.

The tools for computer design of circuits and electronic systems presented in the monograph are intended for use primarily in the Linux operating system. Therefore, the second chapter begins with a concise description of the basic commands in this system necessary to run these tools. The Octave and Matlab environments, which are very useful for solving signal processing tasks, are also briefly discussed. Sections 2.2–2.4 of this chapter are purely mathematical in nature, related to the bilinear transformation between the analog *s* and discrete *z* domains of multidimensional polynomials, and the classes **SNc** and **MDp** created in the C++ language. It has been shown that in the case of multidimensional polynomials, matrix notation is not sufficient and it is necessary to use tensor notation.

The aim of the third chapter is to show the reader the importance of integrated circuit technology for the development of analog signal processing methods. This technology, developed mainly for the fabrication of digital circuits, creates a number of limitations for analog circuits. The main limitation is the low supply voltage, which significantly reduces the possibility of using operational amplifiers (OpAmps) as the basic macrocell in analog systems. This chapter presents two techniques that are least susceptible to such limitations: circuits with OTA-C and current-mode, in particular SI circuits. Based on the square-law FET model, the principle of operation of the basic macrocells for these techniques: SI integrator and OTA was explained.

Chapter 4 shows the application of the techniques presented in Chapter 3. The filter with improved linearity of the phase characteristics is implemented in the OTA-C technique. In turn, two filter banks for the frequency-interleaved analog-to-digital converter (FI ADC), analyzing and synthesizing, are implemented in the switched-current technique (SI) and as a digital filter, respectively. Due to its phase characteristics, the filter implemented in the OTA-C technique is irreversible, while the analyzing and synthesis filters are multiport systems. Their design is possible thanks to the use of the gC prototype circuit. This approach to design has so far only been presented in scientific articles.

Chapter 5 also presents a new approach to video filter design using a prototype multidimensional gC circuit. Hardware implementation on FPGA shows that the multidimensional infinite impulse response (IIR) filter processes real-time video with enhanced quality.

It is assumed that the reader has knowledge of the Fourier transforms $\mathcal{F}$, Laplace $\mathcal{L}$, and discrete $\mathcal{Z}$ in signal processing (Lyons, 2001; Oppenheim et al., 1999), as well as has basic knowledge of electrical circuits described by the laws Ohm and Kirchhoff, composed of R, L, C elements. Elementary knowledge of FETs and integrated circuits may also be useful. The method of using the compiled programs made available to the reader is described in detail in the monograph and does not require advanced IT knowledge. Source version programs are also available. However, for their creative use, deeper IT knowledge is needed, mainly related to programming in the C++ language. The developed programs are available online at http://github.com/AHandkiewicz/IIR3D.

1

Filter Design Based on Lossless Prototype Circuits

From the beginning of the first applications, the filter design was based on ladder LC (inductance, capacitance) structures. Theoretical research on LC circuits was initiated by R. M. Foster (1924). Resistively (R) terminated, they are low sensitive to the changes of the circuit parameters (Orchard, 1966; Orchard et al., 1985). RLC circuits, as passive ones, are stable and moreover, as lossless ones, they are low noise as will be presented in Section 1.4. Such features make the RLC circuits still popular, and they are used as so-called prototype circuits for their counterpart networks fabricated in the form of integrated circuits in OTA-C, SI, and SC techniques. It is assumed that the counterpart circuits realized in the chosen technique inherit the features of the prototype circuits (Handkiewicz, 2002).

In this book the gyrator–capacitor (gC) circuits as the prototype ones are used. The gyrator, introduced by B. D. H. Tellegen (1948), similarly to the inductance and capacitance is a lossless element. In opposition to R, L, and C elements being one-port devices, the gyrator (g) is a two-port network. The symbol of the gyrator is presented in Fig. 1.1. The admittance matrix of the ideal gyrator is as follows:

$$Y = \begin{bmatrix} 0 & g \\ -g & 0 \end{bmatrix}. \tag{1.1}$$

Hence, the dependence of the currents with respect to voltages can be described by the matrix relation

$$I = YV, \tag{1.2}$$

where $I = [I_1 \ \ I_2]'$, $V = [V_1 \ \ V_2]'$, and the symbol $'$ denotes the transposition.

The gC circuits considered in the monograph can be not only two-port networks but also systems with any number of ports. However, the most important feature is multidimensionality – the capacitances of the circuit can be assigned to operators s in different dimensions. For example, in an acoustic filter, as it is a one-dimensional circuit, the time t is assigned to the single operator s by the Laplace transform ($\mathcal{L}$). For the sampled acoustic signal, the discrete time is

Methods and Tools for Computer-aided Design of Multidimensional Filters in Signal Processing,
First Edition. Andrzej Handkiewicz.

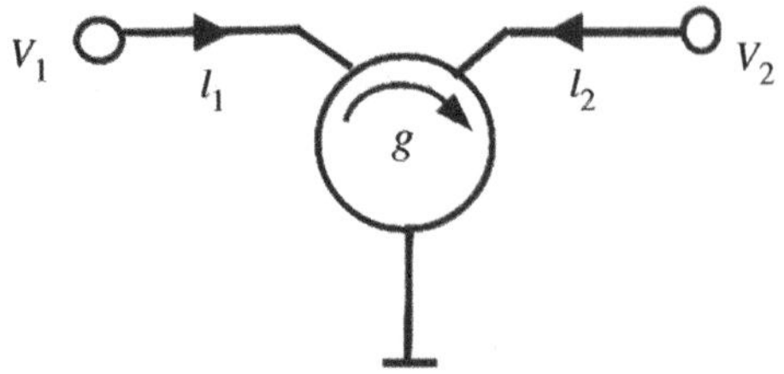

Figure 1.1 Symbol of ideal gyrator with currents and voltages represented on its ports.

assigned to the operator z of the transform $\mathcal{Z}$. The relationship between the s and z domains is given by $z = e^{sT}$, where T is the sampling period. To obtain a description of filters in the frequency domain ω, the equality $s = j\omega$ resulting from the relationship between the Laplace transform $\mathcal{L}$ and the Fourier transform $\mathcal{F}$ is used. In the image filter, being a two-dimensional network, there are two operator variables z_1, z_2 of the $\mathcal{Z}$ transform, corresponding to the discrete variables i_1, i_2, denoting the row and column numbers in the image. In the case of video processing, there is also a third dimension, which is the frame number i_3 in this sequence. The three-dimensional digital equivalent network in the domains z_i, $i = 1, 2, 3$, is derived from the prototype gC circuit in the domains s_i, $i = 1, 2, 3$, using a bilinear transformation (2.3), as the relationship between these variables.

We will start this chapter with a simple example of a third-order elliptic filter. In the diagram, we will present both its ladder LC structure and the corresponding gyrator–capacitor implementation. However, the most important thing is the textual description of this gC circuit in VHDLA. The description of a circuit in this language has a double meaning. On the one hand, it allows you to analyze the more complex networks, for which the graphics representation is not very transparent. On the other hand, this description is a starting point for the symbolic analysis of the designed gC circuit in the **gCstudio** environment. In the resulting file, we will show the effect of such an analysis in the form of so-called structural numbers (SN). Filter design involves calculating the parameters g and C of the circuit for given transfer function coefficients. For this purpose, the **NANOstudio** environment is used with the **optimizer** tool, a detailed description of which can be found in this chapter. Sections 1.2 and 1.3 describe the synthesis, possible in the presented environments, of a seventh-order elliptic filter and a so-called filter pair. In Section 1.4, the balance of powers transferred in the lossless system allows for sensitivity analysis of the considered gyrator–capacitor circuits.

1.1 Design Example of a Simple gC Circuit

A third-order elliptic filter of the LC ladder structure is shown in the upper part and its gC counterpart circuit in the lower part of Fig. 1.2. Both the numerator $N(s)$

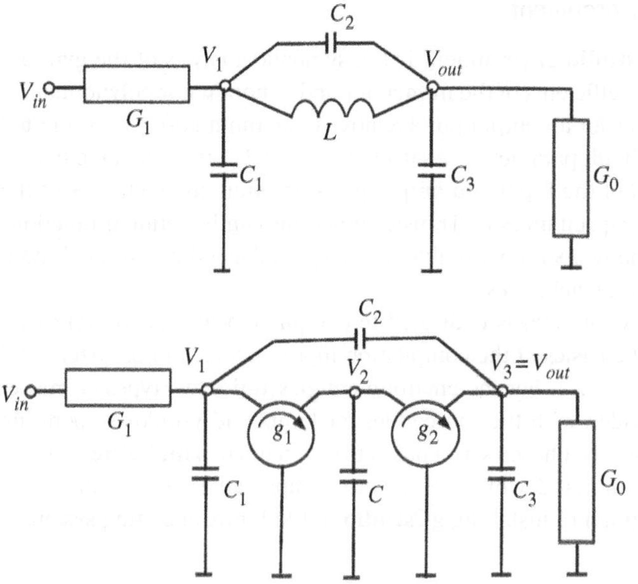

Figure 1.2 Ladder structure of elliptic LC filter and its gC counterpart circuit.

and the denominator $D(s)$ of the transfer function

$$H^a(s) = V_{in}(s)/V_{out}(s) = N(s)/D(s), \tag{1.3}$$

of the filter are third-order polynomials. The coefficients of these polynomials can be obtained in the Matlab or Octave environments using the *ellip* function. Calling, for example, in Octave $[N, D] = ellip(3, 0.5, 30, 1,'s')$, where the arguments of the function *ellip* mean the order (3), the ripple in the passband (0.5 dB), the attenuation in the band-stop (30 dB), and the cut-off frequency (1 rad/s), respectively, gives vectors of the coefficients ordered in descending powers of the *s* variable:

$$N = [0 \quad 0.16902 \quad 0.0000 \quad 0.80290],$$
$$D = [1.0000 \quad 1.22866 \quad 1.51916 \quad 0.80290]. \tag{1.4}$$

By designing the example gC circuit in Fig. 1.2, we mean finding its parameters for given transfer function coefficients. For this purpose, two environments have been developed: **gCstudio** for symbolic analysis of the gC circuit and **NANOstudio**, which enables, among others, solving the system of equations obtained by matching of coefficients presented in symbolic form using **gCstudio**, with the coefficients given in (1.4). We will discuss both of these environments in Sections 1.1.1 and 1.1.2 by illustrating their operation for the considered example.

1.1.1 gCstudio Environment

The goal of the **gCstudio** environment is the symbolic analysis of the gyrator–capacitor circuits. Coefficients of the numerator and denominator polynomials of the transfer functions for all output ports relative to all input ports are computed as functions of gC circuit parameters. As it shown in Fig. 1.2, these parameters are the conductances G of the input and output ports, the transconductances of the gyrators g, and the capacitances C. Transfer functions can be rational functions $H(s_1, s_2, s_3, \dots)$ of many operator variables $s_1, s_2, s_3, \dots$, depending on the dimensions assigned to the capacitances.

The **gCstudio** environment is dedicated to compilation in GCC (GNU Compiler Collection). The basics of the compilation in a Linux operating system will be presented in Section 2.1. For the environment **gCstudio**, two types of **makefile** scripts are provided with the source files for Linux and Windows operating systems, respectively. The chosen script ought to be renamed to **makefile** and executed by the **make** command, which will create a compiled file in the *bin* directory.

A detailed description of installing **gCstudio** in the Linux operating system is as follows:

1. create a shared directory, for example, named *gCstudio*, and move the source subdirectories *lib* and *src* to it,
2. create a subdirectory *bin* in the shared directory,
3. enter the directory *src*: **cd src**,
4. copy makefile_unix : **cp makefile_unix makefile**,
5. execute the **make clean** command – the *.o files in the *src* directory are deleted (this is not necessary for the first compilation),
6. execute the command **make all** – the automatic compilation described in the file **makefile** is executed, and
7. run gc_analyser in the *bin* subdirectory: **./gc_analyser** or, after copying, in any other directory with examples. After starting the program, follow the displayed instructions.

The *gCstudio* directory contains both the compiled program and its source version.

Installation is also possible in the Windows operating system. It is then necessary to use the **MinGW** tools. In the created folder, for example, named *MinGW*, you should:

1. add the path to … \MinGW\bin in system settings,
2. enter the directory *src* in the command window (cmd): **cd src**,
3. copy makefile_win: **copy makefile_win makefile**,
4. run the make command (compilation automation after removing old files): **mingw32-make**,
5. go to *bin* and run the program: **gc_analyser.exe**.

After running the program compiled in Linux with the command **./gc_analyser**, there is displayed a prompt to enter the name of the VHDLA file with a description of the gyrator–capacitor circuit, as it is shown in the following listing:

```
./gc_analyser

GC analyser version: 20150620

 MULTI DIMENSIONAL
______________________________________
Enter VHDLA file name:3ellip1D.vhdla

Options:
        1: compute SN solution
        2: export to Matlab script

Your choice:1

Parsing file 3ellip1D.vhdla ...
8 components found
```

Symbolic analysis is performed when option **1** (compute SN solution) is selected. Option **2** allows to get a Matlab script with matrices whose determinants are polynomials of the transfer function denominator and numerator.

For the circuit whose diagram is shown at the bottom of Fig. 1.2, the file named **3ellip1D.vhdla** is as follows:

```
library VLSI;
use VLSI.SI.all;
entity example is port (
        terminal input  Vin : electrical;
        terminal output Vout : electrical;
        terminal ground gnd : electrical);
end entity example;
-- some comment line
architecture example_arch of example is
variable v1,v2,v3 : real;
begin
-- capacitors with different dimensions
C1: GC_CAP generic map(dim=>1) port map (Ni=>v1, Nj=>gnd);
C: GC_CAP generic map(dim=>1) port map (Ni=>v2, Nj=>gnd);
C3: GC_CAP generic map(dim=>1) port map (Ni=>v3, Nj=>gnd);
C2: GC_CAP generic map(dim=>1) port map (Ni=>v1, Nj=>v3);
-- conductances with I/O signals
Gi: GC_CON port map (Ni=>Vin, Nj=>v1);
Go: GC_CON port map (Ni=>v3, Nj=>gnd, Vo=>Vout);
-- gyrators
ig1: GC_GYR port map( Ni=>v1, Nj=>v2);
ig2: GC_GYR port map( Ni=>v2, Nj=>v3);

end architecture;
```

According to the *VHDLA* language syntax, terminals are declared in the file header. In the body of the architecture, they are declared, as the variables, the node voltages of the circuit, and then all circuit elements are listed. The order of the element's listing is arbitrary. The name of the capacitance begins with the letter **C**, followed by the dimension assigned to the capacitance and the nodes between which it is connected. All of the ideal gyrators are described by sequences of characters beginning from the letters **ig**, and the load conductances begin from the letter **G**.

The results of running **./gc_analyser** for option **1** are placed in two files: **3ellip1D.v.sn** and **CircuitElements**. The first one contains symbolic coefficients of the denominator and numerator polynomials:

```
%  Ci  Cj  Cf  Ig  Gi1  Gi2  Gj1  Gj2  Ni  Nj
%! BEGIN DEN
%! NAME DEN
%! TYPE 2
%! IN #
%! OUT #
%! RANK 3
%! DIMENSION 1
%! DRP 3
%! LENTRIES 15
%! COLUMNS 3
%-----------------
% S 3
%-----------------
   1   1   2   3
   1   1   2   4
   1   2   3   4
%-----------------
% S 2
%-----------------
   1   1   2   8
   1   2   3   7
   1   2   4   7
   1   2   4   8
%-----------------
% S 1
%-----------------
   1   2   7   8
   1   1   6   6
   1   4   6   6
   1   3   5   5
   1   4   5   5
  -2   4   5   6
%-----------------
% S 0
%-----------------
   1   6   6   7
   1   5   5   8
```

```
%---------------
%! END DEN
%! BEGIN VinVout
%! NAME VinVout
%! TYPE 1
%! IN Vin
%! OUT Vout
%! RANK 2
%! DIMENSION 1
%! DRP 2
%! LENTRIES 2
%! COLUMNS 3
%---------------
% S 2
%---------------
  1   2   4   7
%---------------
% S 0
%---------------
  1   5   6   7
%---------------
%! END VinVout
```

The first part of the above file, between the **BEGIN DEN** and **END DEN** lines correspond to the denominator of the transfer function (1.3). The second part, between the **BEGIN VinVout** and **END VinVout** lines, corresponds to the numerator of the transfer function (1.3). Such symbolic representations of these polynomials we will call SN (Bellert, 1962). In the file **3ellip1D.v.sn**, there are two SN the first corresponding to the denominator and the second to the numerator of the transfer function. They are composed of rows, each containing the same number of digits, the first of which is a coefficient. The remaining digits, in conjunction with the **x** variable, represent the parameters of the circuit, as described in the second file **CircuitElements**:

```
Structure Number calculated for the circuit:
Ci    C1    alias x1:    sign=1 1    0    dim=1
Ci    C     alias x2:    sign=1 2    0    dim=1
Ci    C3    alias x3:    sign=1 3    0    dim=1
Cf    C2    alias x4:    sign=1 1    3    dim=1
Ig    ig1   alias x5:    sign=1 1    2
Ig    ig2   alias x6:    sign=1 2    3
Gi1   Gi    alias x7:    sign=1 0    1    signal=Vin
Gj2   Go    alias x8:    sign=1 3    0    signal=Vout
Analysing admittance matrix:               DONE
Calculating determinant:                   DONE
Analysing signal path from Vin to Vout:    DONE
Calculating determinant:                   DONE
Finished in 0 second(s).

Results were saved as: 3ellip1D.v.sn
Finished in 0 seconds
```

For example, the capacitance **C1** in Fig. 1.2 is marked as **x1**, and in SN, both the denominator and the numerator, are described by the digit **1**. In the case of **C2**, this digit is **4**. Both capacitances are assigned to the first dimension. SN are calculated from the ***.sn** file using the members of the **SNc** (SN calculator) class, described in detail in Chapter 2. SN are divided into subsets, separated by dashed lines in the file ***.sn**, and assigned to powers of the complex variable s. For example, in the above file **3ellip1D.v.sn**, the denominator of the transfer function has the SN composed of 15 lines (*LENTRIES 15* in the description header) divided into four subsets. The subsets correspond to the complex variables s^3, s^2, s^1, s^0, with the exponents given in the lines starting with *% S3*, *% S2*, *% S1*, *% S0*, respectively. For example, after the line *% S3* there is a subset denoting the symbolic form of the polynomial coefficient assigned to the complex variable s^3. Hence, the corresponding term of the polynomial can be written explicitly as $s^3(1 \cdot x1 \cdot x2 \cdot x3 + 1 \cdot x1 \cdot x2 \cdot x4 + 1 \cdot x2 \cdot x3 \cdot x4)$.

As mentioned before, option **2** in the **gCstudio** environment produces the matrices of the gC circuit node equations, saved in the file **3ellip1D.v.m** as follows:

```
% Ci  Cj  Cf  Ig  Gi1  Gi2  Gj1  Gj2  Ni  Nj
syms s x1 x2 x3 x4 x5 x6 x7 x8 ;
% ------------------------------------------------
% Admittance matrix
% ------------------------------------------------
Y =[x1*s+x4*s+x7,x5       ,-x4*s  ;
        -x5       ,x2*s     ,x6     ;
        -x4*s     ,-x6      ,x3*s+x4*s+x8]
% ------------------------------------------------
%INPUT Vin versus OUTPUT Vout
% ------------------------------------------------
Vin_Vout =[      x1*s+x4*s+x7,x5        ,x7       ;
        -x5       ,x2*s     ,0        ;
        -x4*s     ,-x6      ,0        ]
% ------------------------------------------------
```

The determinants of these matrices are the polynomials of the denominator $D(s)$ and numerator $N(s)$ of the transfer function (1.3) of the circuit in Fig. 1.2. The symbolic analysis algorithm in the **gCstudio** environment is based on calculating these determinants.

The last stage of the filter synthesis is solving the system of equations obtained by coefficient matching. This step is accomplished by the **MDp** class, (multidimensional polynomial), also described in Chapter 2. In the presented example, such a system of equations is as follows:

$$
\begin{aligned}
1 \cdot x1 \cdot x2 \cdot x3 + 1 \cdot x1 \cdot x2 \cdot x4 + 1 \cdot x2 \cdot x3 \cdot x4 = 1,\\
1 \cdot x1 \cdot x2 \cdot x8 + 1 \cdot x2 \cdot x3 \cdot x7 + 1 \cdot x2 \cdot x4 \cdot x7 + 1 \cdot x2 \cdot x4 \cdot x8 = 1.22866,\\
1 \cdot x2 \cdot x7 \cdot x8 + 1 \cdot x1 \cdot x6 \cdot x6 + 1 \cdot x4 \cdot x6 \cdot x6 + 1 \cdot x3 \cdot x5 \cdot x5 +\\
+1 \cdot x4 \cdot x5 \cdot x5 - 2 \cdot x4 \cdot x5 \cdot x6 = 1.51916,
\end{aligned}
$$

$$\begin{aligned}1 \cdot x6 \cdot x6 \cdot x7 + 1 \cdot x5 \cdot x5 \cdot x8 &= 0.80290,\\ 1 \cdot x2 \cdot x4 \cdot x7 &= 0.16902,\\ 1 \cdot x5 \cdot x6 \cdot x7 &= 0.80290.\end{aligned} \tag{1.5}$$

Note that the symbolic figures received have a special form. Each of them is the sum of the products of the same number of variables equal to the order of the filter. This means that the parameters of the circuit obtained as a solution to such a system of equations represent the transfer function (1.3) also after k times scaling. In the example above, there are three variables in each factor of the transfer function numerator and denominator polynomials. After scaling, the polynomial coefficients will change k^3 times, which means that after dividing these polynomials by k^3, the transfer function itself does not change. For numerical reasons, this is an important feature that allows us to search for **normalized** parameter values, e.g. capacitance about 1 F instead about 1 μF or conductance about 1 S instead of about 1 μS. We assume, therefore, that the transfer function is also given in the **normalized** form, i.e. the filter limit frequencies are of the order of 1 rad/s. We determine the values of the parameters, being the solution of equation (1.5), using the environment presented in Section 1.1.2.

1.1.2 NANOstudio Environment

The parameters of the gC prototype circuit in Fig. 1.2 are calculated in the **NANOstudio** environment using the optimization method for a randomly generated starting point. As mentioned before, equation (1.5) have a special symbolic form. For this reason, the Hooke–Jeves algorithm turns out to be fast and effective. This gradientless algorithm, first presented in 1961, is still used in many modern applications (Altinoz and Yilmaz, 2015; Braun et al., 2015; Wyers et al., 2016). The presented version of the algorithm can be modified for discrete problems and was presented in earlier works on technological migration (Naumowicz et al., 2013) and optimization (Sniatala et al., 2015).

The **NANOstudio** source files are located in the *src* directory. In addition to the main program **hooke-jeeves.c** the directory contains the subdirectory *includes* the library files **SNc.h** and **MDp.h** containing the aforementioned classes. The classes are used to generate the SN and the multidimensional matrices on the basis of the file ***.sn**, calculated in **gCstudio** environment. Compilation can be made after placing the script **makefile** into the directory created by the user. Thanks to the script, the compilation in the Linux operating system is automatically made by the instruction **make all** invoked in the created directory, in subsequent compilations preceded by the instruction **make clean**.

The program **optimizer** created in this way can be used with the following options:

-r – optimization with respect to the parameters of the gC multiport circuit in the range of real numbers without constraints,
-x – optimization with respect to the parameters of the gC multiport circuit in the range of real numbers with constraints,
*-c*1 – storing the results in memory (1 means no storing),
-a – design with a specific type of analysis: in the analog domain for the parameter *s* or, after the bilinear transformation, in the complex domain for the parameter *z* and in the discrete time domain for the parameters *t* or *m*,
-n – generating random variables, in a given range (option *-x*) or without constraints,
-i – number of iterations performed during optimization,
-t – the initial value of the variable step (τ) at the start of the algorithm,
-d – number of samples in discrete time for particular dimensions of the gC circuit. Discrete time can also be interpreted as a row or a column number, or a frame number in a video.

The type of analysis in complex domains or in the discrete time domain is implemented for the synthesized gC multiport circuit by specifying the appropriate parameter for the option *-a*:

-a s – analysis in *s* domain; it is assumed that the files *s*0, *s*1, ... that correspond to the given transfer functions, are placed in the working directory: *s*0 is the file of the transfer function denominator coefficients, *s*1, *s*2, ... are the files of the transfer function numerator coefficients,
-a z – analysis in *z* domain; it is assumed that the files *z*0, *z*1, ... that correspond to the given transfer functions, are placed in the working directory: *z*0 is the file of the transfer function denominator coefficients, *z*1, *z*2, ... are the files of the transfer function numerator coefficients,
-a t – discrete time analysis *t*; it is assumed that the files *t*0, *t*1, *t*2, ... that correspond to the samples on the all input ports (*t*0) or on the output ports 1, 2, ..., respectively, are given,
-a m – discrete time analysis with optimization for input samples included in the given file *t*0 and for the response samples included in the given file *t*1.

Note: Options *s* and *z* concern the synthesis of the multiport gC circuit in its complex domains, while options *t* and *m* concern the synthesis in its discrete-time domains. For options *z*, *t*, and *m*, a bilinear transformation is used to replace the domain *s*, in which the prototype gC circuit is described, with the domain *z* for its discrete counterpart network, as it is shown in Section 2.3. The options *t* and *m* concern the analysis in the domain of discrete time with the multiplication of transfer function polynomials of two systems: one of them described by the designed prototype multiport gC circuit, and the second network described by the denominator and numerator matrices $zd1, zn1, zd2, zn2, \ldots$ of the transfer functions in the domain z, respectively. These transfer functions allow taking into account, already at the stage of synthesis of the prototype gC circuit, the cooperation of its discrete counterpart with the digital circuit described by these transfer functions. The matrices $zd1, zn1, zd2, zn2, \ldots$ of the given transfer functions may in particular be unitary. The m option is for a multiport network with one input and many outputs connected to circuits with given transfer functions. The t option is for a multiport network with one output and many inputs connected to a common input by circuits with given transfer functions. The number of samples for t and m parameters are given, for the appropriate dimension, as parameters of the d option. It is worth adding that each user can construct his own options by making appropriate changes in the programs **main.c** and **sn.cpp**.

According to the options above, the execution of the compiled **optimizer** in the for-loop

```
for ((i=0; i<=9; i++));
do
./optimizer -c 1 -x -n -t 0.001 -a s -i 1000 3ellip1D.sn init > "f$i"
grep Goal "f$i"
done
```

allows to find the values of the parameters of the prototype gC circuit with the structure **3ellip1D.vhdl**, presented in Section 1.1.1. The solutions obtained in the next steps are placed into the files fi, $i = 0, \ldots, 9$ using the shift operator >. The **grep** command allows to monitor the progress in the calculations. The file **init** is in the form:

```
0.5 0 1 0.000001
0.5 0 1 0.000001
0.5 0 1 0.000001
0.5 0 1 0.000001
0  -1 1 0.000001
0  -1 1 0.000001
0.5 0 1 0.000001
0.5 0 1 0.000001
```

meaning that for the capacitances, represented by the first four lines, and the load conductances, represented by the last two lines, the searching for solutions in the range from 0 to 1 (the 2nd and 3rd numbers in each line) was adopted. The remaining lines in the above printout mean that for the gyrator conductances, the searching for solutions in the range from −1 to 1 was assumed. The 1st number (0 or 0.5) in each line means the initial value of the circuit parameter, if it is not randomly generated, and the last one (0.000001) is assumed calculation accuracy.

The designer decides on the choice of solution, guided mainly by the value of the objective function saved together with the values of the circuit parameters in the files fi, $i = 0, \dots, 9$. For example, for the objective function with a value of 0.00000000000006, the following parameter values were achieved

```
0.740619838840654
0.999423843348229
0.961527141230247
0.169462664984342
-0.959979066177258
-0.838080235881442
0.997962314578629
0.110630682694843
```

corresponding to the elements of the gC circuit, as described in **CircuitElements**.

According to the data in (1.4), the files $s0$ and $s1$ are of the form:

```
1
3
0 0.80290
1 1.51916
2 1.22866
3 1
```

and

```
1
3
0 0.80290
1 0
2 0.16902
3 0,
```

respectively.

In addition to the transfer function coefficients, these files contain the circuit dimension (the number 1 in the first line) and the order of the numerator and the denominator (the number 3 in the second line). In the following lines, the first number denotes the exponent of the power of the s variable and the second value of the coefficient at s in this power. If any of the coefficients is zero, the corresponding line can be omitted.

1.2 Higher Order Filters

For both the elliptic filter and the filters designed by Butterworth and Chebyshev approximations, their transfer functions can be obtained, both in the s and z domains, as kth degree rational functions, which we will write as follows:

$$H^a(s) = \frac{P'S}{Q'S}, \tag{1.6}$$

where P and Q are the vectors of the numerator and denominator coefficients of the transfer function, and S the vector containing the complex variables s in the appropriate powers:

$$S' = [s^0 \; s^1 \; s^2 \; s^3 \ldots s^k]. \tag{1.7}$$

The symbol $'$ in the above equations means vector (matrix) transposition. The notation of the S vector according to the powers of the variable s ascending, unusual, for example, for Matlab or Octave environments based on matrix calculus, will find its justification in Section 2.3. In the general case, a multiport circuit of any dimension, it will be necessary to use tensor notation instead of matrix notation.

The representation of the transfer function in the discrete domain z is similar to that in the analog domain:

$$H(z) = \frac{A'Z}{B'Z}, \tag{1.8}$$

where the vectors of the transfer function polynomial coefficient are now denoted A and B, while Z is the vector containing the complex variables z^{-1} in the appropriate powers:

$$Z' = [z^0 \; z^{-1} \; z^{-2} \; z^{-3} \ldots z^{-k}]. \tag{1.9}$$

In this section, we will synthesize a seventh-order elliptic filter using the **gCstudio** and **NANOstudio** environments. The given transfer function coefficients entered into the file *s0*

```
1
7
0 0.10726
1 0.49362
2 1.08843
3 2.12027
4 2.12978
5 2.64085
6 1.14091
7 1
```

and into the file *s1*

```
1
7
0 0.10726
1 0
2 0.14345
3 0
4 0.05802
5 0
6 0.00659
7 0
```

respectively, were obtained using the function $[N, D] = ellip(7,\ 0.5,\ 60,\ 1,\text{'s'})$ in Octave.

The structure of the synthesized filter has the following description in VHDLA

```
library VLSI;
use VLSI.SI.all;
entity example is port (
        terminal input Vin : electrical;
        terminal output Vout : electrical;
        terminal ground gnd : electrical);
end entity example;
-- some comment line
architecture example_arch of example is
variable v1,v2,v3,v4,v5,v6,v7 : real;
begin
-- capacitors with different dimensions
C1: GC_CAP generic map(dim=>1) port map (Ni=>v1, Nj=>gnd);
C2: GC_CAP generic map(dim=>1) port map (Ni=>v2, Nj=>gnd);
C3: GC_CAP generic map(dim=>1) port map (Ni=>v3, Nj=>gnd);
C4: GC_CAP generic map(dim=>1) port map (Ni=>v4, Nj=>gnd);
C5: GC_CAP generic map(dim=>1) port map (Ni=>v5, Nj=>gnd);
C6: GC_CAP generic map(dim=>1) port map (Ni=>v6, Nj=>gnd);
C7: GC_CAP generic map(dim=>1) port map (Ni=>v7, Nj=>gnd);
Cp1: GC_CAP generic map(dim=>1) port map (Ni=>v1, Nj=>v3);
Cp2: GC_CAP generic map(dim=>1) port map (Ni=>v3, Nj=>v5);
Cp3: GC_CAP generic map(dim=>1) port map (Ni=>v5, Nj=>v7);
-- conductances with I/O signals
G1: GC_CON port map (Ni=>Vin, Nj=>v1);
G2: GC_CON port map (Ni=>v7, Nj=>gnd, Vo=>Vout);
-- gyrators
ig1: GC_GYR port map( Ni=>v1, Nj=>v2);
ig2: GC_GYR port map( Ni=>v2, Nj=>v3);
ig3: GC_GYR port map( Ni=>v3, Nj=>v4);
ig4: GC_GYR port map( Ni=>v4, Nj=>v5);
ig5: GC_GYR port map( Ni=>v5, Nj=>v6);
ig6: GC_GYR port map( Ni=>v6, Nj=>v7);

end architecture;
```

The result file **7ellip1D.sn**, of symbolic analysis carrying out in **gCstudio** environment as described in Section 1.1.1, is not shown due to its size. The file **init**, created in the same way as in Section 1.1.2, has the following form:

```
0.5 0 1.5 0.000001
0.5 0 1.5 0.000001
0.5 0 1.5 0.000001
0.5 0 1.5 0.000001
```

```
0.5 0 1.5 0.000001
0.5 0 1.5 0.000001
0.5 0 1.5 0.000001
0.5 0 1.5 0.000001
0.5 0 1.5 0.000001
0.5 0 1.5 0.000001
0 -1.5 1.5 0.000001
0 -1.5 1.5 0.000001
0 -1.5 1.5 0.000001
0 -1.5 1.5 0.000001
0 -1.5 1.5 0.000001
0 -1.5 1.5 0.000001
0.5 0 1.5 0.000001
0.5 0 1.5 0.000001
```

We find the values of the gC circuit parameters by running the optimization program

```
for ((i=0; i<=9; i++));
    do
./optimizer -c 1 -x -n -t 0.00001 -a s -i 100000 7ellip1D.sn init
                                                          > "f$i"

    grep Goal "f$i"
    done
```

One of the solutions, obtained for the value of the objective function equal to 0.000000285802377, is

```
0.859881293321741
0.692049326580107
1.499997146076000
0.338611251260261
1.356699841360094
0.701048514347098
0.760056100407697
0.847087556576119
0.419787131389294
0.091998201779113
1.007539728604105
0.938498463790234
0.550937714150477
0.482560344512586
0.650425411871593
0.545190265328167
1.203570600918757
0.257999309128753
```

for which the frequency response is shown as a dotted line in Fig. 1.3.

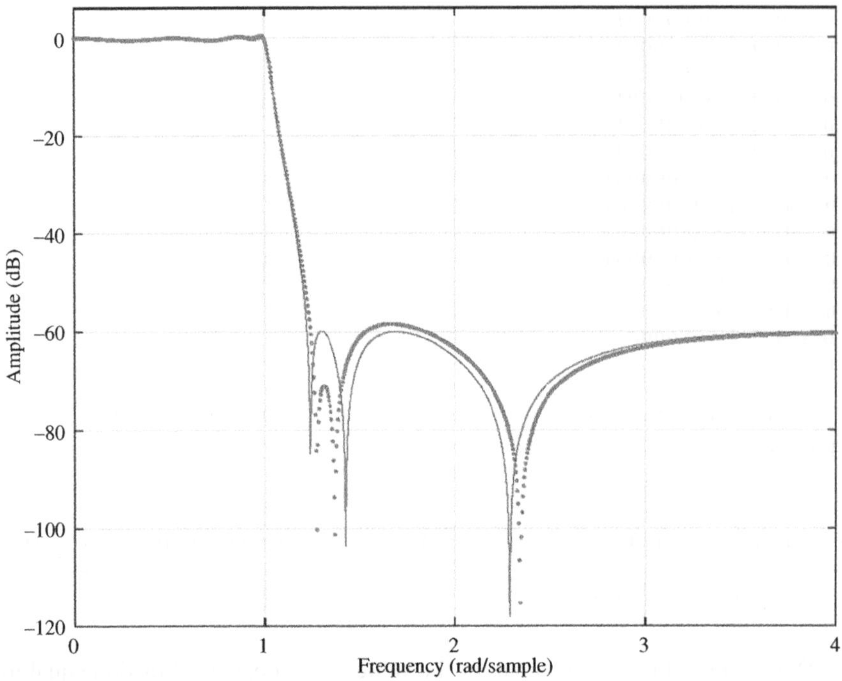

Figure 1.3 Frequency characteristics of 7th order elliptic filter calculated for coefficients obtained in Octave (solid line) and after gC circuit synthesis (dotted line).

1.3 Filter Pair Design

In the case of arbitrary multiport networks, there are no such design methods as exist for two-port ones, allowing to obtain the transfer functions in the complex domain s of Butterworth, Chebyshev and elliptic filters, respectively. The exception is the so-called filter pair that is the multiport circuit with one input and two outputs, or vice versa, with two inputs and one output. In this section, we will show how the **gCstudio** and **NANOstudio** environments work for such a case (Handkiewicz et al., 2012).

We will write the transfer functions of a filter pair similarly to (1.6) as

$$H_l(s) = \frac{P_l' S}{Q' S}, \tag{1.10}$$

and

$$H_h(s) = \frac{P_h' S}{Q' S}, \tag{1.11}$$

which can be described by

$$H_l(s) = \left[\frac{D_o(-s)}{D_o(s)} + \frac{D_e(-s)}{D_e(s)}\right]/2, \tag{1.12}$$

and

$$H_h(s) = \left[\frac{D_o(-s)}{D_o(s)} - \frac{D_e(-s)}{D_e(s)}\right]/2, \tag{1.13}$$

where $D_o(-s)/D_o(s)$, $D_e(-s)/D_e(s)$ correspond to stable all-pass filter transfer functions obtained on the basis of odd and even parts of the denominator polynomials of a ladder LC filter (Willson and Orchard, 1995; Saramaki, 1985; Handkiewicz and Naumowicz, 2021).

The power balance on the filter pair ports implies the following relationship of magnitude responses:

$$|H_l(j\omega)|^2 + |H_h(j\omega)|^2 = 1, \tag{1.14}$$

(Läwenborg et al., 2003). Hence, unlike single filters, in the filter pairs the stopband attenuation a_s and the passband ripple a_p are dependent on each other, according to the formula:

$$\frac{1}{a_p^2} + \frac{1}{a_s^2} = 1, \tag{1.15}$$

which in nepers gives

$$e^{-2a_p} + e^{-2a_s} = 1, \tag{1.16}$$

($a_{Np} = ln(a)$), (Willson and Orchard, 1995), or in *dB*

$$10^{-0.1a_p} + 10^{-0.1a_s} = 1, \tag{1.17}$$

($a_{dB} = 20\log(a)$). To calculate in Octave the transfer functions according to (1.12) and (1.13) the following function can be used:

```
function retVal=FiltPair(n,as)
%Transfer functions of the n-th order filter pair
if ((mod(n,2) == 0) || as < 0)
printf("   n must be odd and\n  as must be positive\n")
else
ap=-10*log10(1-10^(-0.1*as));
%Numerator Pl for low-pass output and denomiantor
[Pld,Qd]=ellip(n,ap,as,1,'s');
%Poles calculation
Rs=roots(Qd);
[M, iR]=mpoles(Rs);
for i = 1:n
R(i)=Rs(iR(i));
end
%Alternating distribution of poles
```

```
i=j=k=1;
Ro(j)=R(i); j++;
while (i < n) i++;
if (mod(i,4) == 0)
  Ro(j)=R(i); j++; i++;
  Ro(j)=R(i); j++;
else
  Re(k)=R(i); k++; i++;
  Re(k)=R(i); k++;
end
end
%Denominators of all-pass transfer functions
Do=poly(Ro);
De=poly(Re);
%Numerators of all-pass transfer functions
if (mod((n-1),4) == 0)
n1=(n-1)/2+1; n2=(n-1)/2+2;
else
n2=(n-1)/2+1; n1=(n-1)/2+2;
end
for i=1:n1
D_e(i)=(-1)^(i+1)*De(i);
end
for i=1:n2
D_o(i)=(-1)^(i)*Do(i);
end
%Numerator Ph for high-pass output
Phd=(conv(Do,D_e)-conv(D_o,De))/2;

printf("\n The polynomials in ascending powers of s: \n");
Pl=Ph=Q=zeros(1,n+1);
for i=1:n+1
if (n+1-i > 0) Pl(i)=real(Pld(n+1-i)); end;
Ph(i)=real(Phd(n+2-i));
Q(i)=real(Qd(n+2-i));
end
Q
Pl
Ph

%frequency responses
%w=logspace(0.0001,0.1,248);
w=linspace(0,4,512);
%lowpass output
Hl=freqs(Pld,Qd,w);
%highpass output
Hh=freqs(Phd,Qd,w);

figure 1
%plot(w,abs(Hl),"-",w,abs(Hh),"--");
plot(w,abs(Hl),w,abs(Hh));
```

```
axis([0 4 0 1.1]);
xlabel("Frequency [rad/sample]");
ylabel("Amplitude responses");
grid on
figure 2
plot(w,20*log10(abs(Hl)),"-",w,20*log10(abs(Hh)),".");
axis([0 4 -60 4]);
xlabel("Frequency [rad/sample]");
ylabel("Amplitude [dB]");
grid on
retVal=ap;
end
end
```

in which $D_o(s)$ and $D_e(s)$ are the odd and even parts of the denominator polynomial $Q'S$ in an elliptic low-pass transfer function (1.10). The even and odd parts are calculated after alternating distribution of the polynomial $Q'S$ roots (Gazsi, 1985).

Calling the function *FiltPair*(5, 18) we obtain the example polynomials for the fifth-order pair as follows:

$$Q' = [1.2613 \;\; 2.0935 \;\; 3.2192 \;\; 3.0720 \;\; 1.8241 \;\; 1],$$
$$P'_l = [1.2613 \;\; 0 \;\; 1.5954 \;\; 0 \;\; 0.4685 \;\; 0],$$
$$P'_h = [0 \;\; 0.53503 \;\; 0 \;\; 1.51804 \;\; 0 \;\; 1], \tag{1.18}$$

for the polynomials of the denominator Q and the numerators of the low-pass P_l and high-pass P_h outputs, respectively. The frequency characteristics of the example filter pair are shown in Fig. 1.4.

The structure of the implemented filter pair, composed of 5 grounded capacitances, 10 gyrators, 1 input conductance, and 2 output conductances, has the following description in the VHDLA language:

```
library VLSI;
use VLSI.SI.all;
entity example is port (
        terminal input Vin : electrical;
        terminal output VoutL : electrical;
        terminal output VoutH : electrical;
        terminal ground gnd : electrical);
end entity example;
-- some comment line
architecture example_arch of example is
variable v1,v2,v3,v4,v5,v6,v7: real;
begin
-- capacitors with different dimensions
C1 : GC_CAP generic map(dim=>1) port map (Ni=>v2, Nj=>gnd);
C2 : GC_CAP generic map(dim=>1) port map (Ni=>v6, Nj=>gnd);
C3 : GC_CAP generic map(dim=>1) port map (Ni=>v3, Nj=>gnd);
```

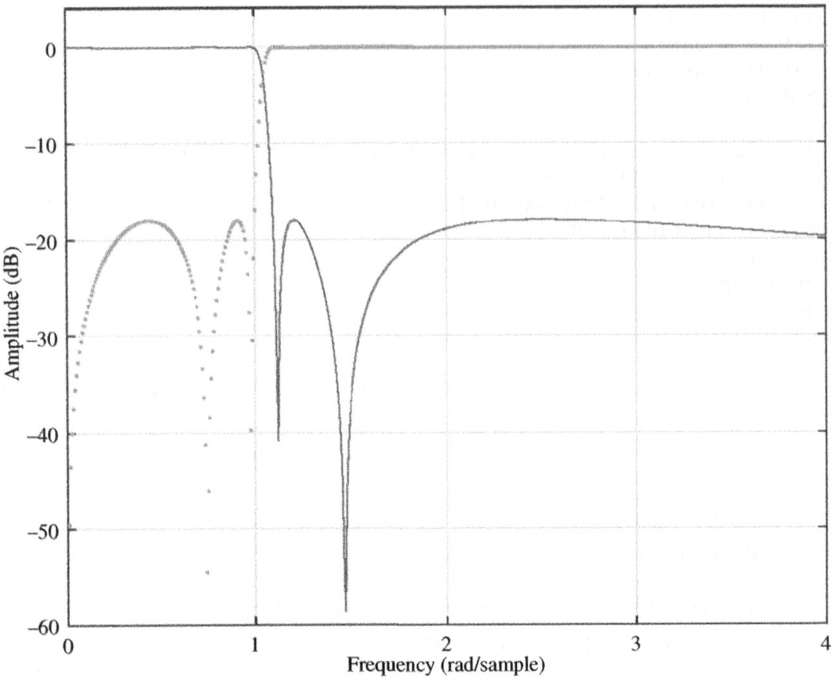

Figure 1.4 Frequency characteristics of a fifth-order filter pair for the low-pass (solid line) and the high-pass (dotted line) outputs, respectively.

```
C4 : GC_CAP generic map(dim=>1) port map (Ni=>v4, Nj=>gnd);
C5 : GC_CAP generic map(dim=>1) port map (Ni=>v5, Nj=>gnd);
-- conductances with I/O signals
G1 : GC_CON port map (Ni=>Vin, Nj=>v1);
G2 : GC_CON port map (Ni=>v6, Nj=>gnd, Vo=>VoutL);
G4 : GC_CON port map (Ni=>v7, Nj=>gnd, Vo=>VoutH);
-- gyrators
ig1 : GC_GYR port map( Ni=>v1, Nj=>v2);
ig2 : GC_GYR port map( Ni=>v1, Nj=>v6);
ig3 : GC_GYR port map( Ni=>v1, Nj=>v7);
ig4 : GC_GYR port map( Ni=>v2, Nj=>v3);
ig5 : GC_GYR port map( Ni=>v3, Nj=>v4);
ig6 : GC_GYR port map( Ni=>v4, Nj=>v5);
ig7 : GC_GYR port map( Ni=>v2, Nj=>v6);
ig8 : GC_GYR port map( Ni=>v2, Nj=>v7);
ig9 : GC_GYR port map( Ni=>v3, Nj=>v6);
ig10 : GC_GYR port map( Ni=>v5, Nj=>v6);

end architecture;
```

while files $s0, s1, s2$, with coefficients set in (1.18), are:

```
1
5
0  1.2613
1  2.0935
2  3.2192
3  3.0720
4  1.8241
5  1

1
5
0  1.2613
1  0
2  1.5954
3  0
4  0.4685
5  0

1
5
0  0
1  0.53503
2  0
3  1.51804
4  0
5  1
```

respectively.

For the file **init**

```
0.5 0 1.5 0.0000001
0.5 0 1.5 0.0000001
0.5 0 1.5 0.0000001
0.5 0 1.5 0.0000001
0.5 0 1.5 0.0000001
0 -1 1.5 0.0000001
0 -1 1.5 0.0000001
0 -1 1.5 0.0000001
0 -1 1.5 0.0000001
0 -1 1.5 0.0000001
0 -1 1.5 0.0000001
0 -1 1.5 0.0000001
0 -1 1.5 0.0000001
0 -1 1.5 0.0000001
0 -1 1.5 0.0000001
0.5 0 1.5 0.0000001
0.5 0 1.5 0.0000001
0.5 0 1.5 0.0000001
```

and after running the program **optimizer**

```
for ((i=0; i<=9; i++)); do
./optimizer -c 1 -x -n -t 0.000001 -a s -i 10000 5pair1Dmin.sn init
                                                              > "f$i"
grep Goal "f$i"
done
```

we obtain as the one of the solutions:

```
1.480492423591296
0.394223370235035
1.492852087227228
1.158425712943560
0.676232696895387
1.355491138990725
0.533177105606084
0.978763762262115
1.335262492984787
0.367545986704895
-0.851097023262463
-0.315513788927167
-0.885928409828217
0.359885237033817
-0.266013396463995
1.496920996046349
0.326702980233175
0.338778324257805
```

reaching the objective function with a value equal to 0.000000001437.

In the file above, as in the filter synthesis in the previous examples, the first 5 numbers represent the capacitance values:

$$C_1 = 1.480492423591296,$$
$$C_2 = 0.394223370235035,$$
$$C_3 = 1.492852087227228,$$
$$C_4 = 1.158425712943560,$$
$$C_5 = 0.676232696895387, \tag{1.19}$$

in Farads (F). The next 10 numbers are the transconductance values of the gyrators found in the circuit description:

$$g_1 = 1.355491138990725,$$
$$g_2 = 0.533177105606084,$$
$$g_3 = 0.978763762262115,$$
$$g_4 = 1.335262492984787,$$
$$g_5 = 0.367545986704895,$$

$$g_6 = -0.851097023262463,$$
$$g_7 = -0.315513788927167,$$
$$g_8 = -0.885928409828217,$$
$$g_9 = 0.359885237033817,$$
$$g_{10} = -0.266013396463995, \tag{1.20}$$

in Siemens (S). Negative transconductance values mean that the gyrators in the circuit are directed opposite to that in the gC circuit description. The last 3 numbers are the input and two output conductance values:

$$G_1 = 1.496920996046349,$$
$$G_2 = 0.326702980233175,$$
$$G_3 = 0.338778324257805, \tag{1.21}$$

also in Siemens (S).

The admittance matrix (1.1) shows that the gyrator can be implemented using two operational transconductance amplifiers (OTA). The designed filter pair can therefore be directly fabricated as an integrated circuit in the OTA-C technique as will be shown in Section 4.1. In addition to digital, another possibility of implementing the gC circuit is the SI technique, as will be shown in the case of a filter bank in Section 4.5.

1.4 Low-Sensitivity Filter Design Strategy

In this section, we will present the basics of the low-sensitive strategy of filter design (Handkiewicz, 2002). We will describe sensitivity properties of a lossless two-port network terminated by resistors, considered as a special case of a multiport network. The lossless two-port network can be not only an LC ladder circuit, but in general a gC circuit of any structure, presented for example in Section 1.3.

As it was shown in Sections 1.2 and 1.3, the voltage transfer function of a doubly loaded lossless two-port network presented in Fig. 1.5, can be written as a rational function in the form

$$H^a(s) = \frac{V_2(s)}{V_{in}(s)} = \frac{N(s)}{D(s)}. \tag{1.22}$$

Assuming that the voltage source, the two-port network, and the load are all matched to each other, the maximum power delivered from the source is

$$P_{max} = \frac{|V_{in}(j\omega)|^2}{4R_1} = \frac{|V_{in}(j\omega)|^2 G_1}{4}. \tag{1.23}$$

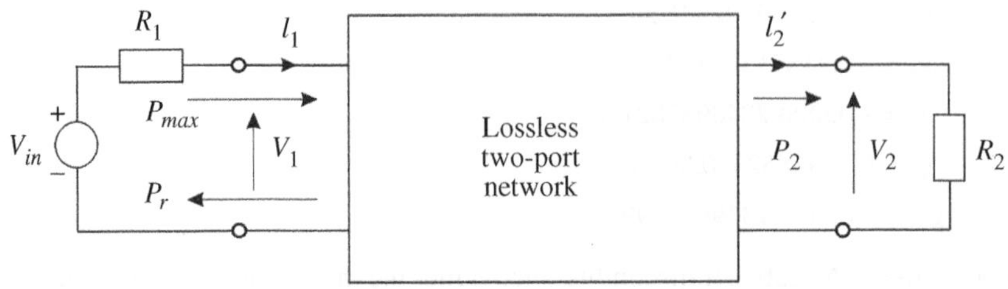

Figure 1.5 Lossless two-port network loaded by resistors.

If the matching conditions are fulfilled, the power dissipated in the load of the lossless two-port network

$$P_2 = \frac{|V_2(j\omega)|^2}{R_2} = |V_2(j\omega)|^2 G_2, \tag{1.24}$$

is equal to P_{max} ($P_2 = P_{max}$). For nominal values of parameters, the two-port network can be matched to its terminations at a source frequency ω_0. We assume that at least one such frequency exists. If matching conditions are not fulfilled because of frequency changes or circuit parameter variations, we have

$$\frac{P_{max}}{P_2} \geq 1, \tag{1.25}$$

or

$$P_{max} = P_2 + P_r, \tag{1.26}$$

where P_r is reflected power. P_r is nonzero when the matching conditions are not fulfilled. Introducing the transducer factor

$$H_t(s) = \frac{1}{2}\sqrt{\frac{G_1}{G_2}\frac{1}{H^a(s)}} = \frac{D(s)}{P(s)}, \tag{1.27}$$

where $P(s) = 2\sqrt{\frac{G_2}{G_1}}N(s)$, we obtain

$$|H_t(j\omega)|^2 = \frac{P_{max}}{P_2}. \tag{1.28}$$

Similarly, we can introduce the characteristic function

$$K(s) = \frac{F(s)}{P(s)}, \tag{1.29}$$

so that

$$|K(j\omega)|^2 = \frac{P_r}{P_2}. \tag{1.30}$$

From relation (1.26) we obtain the Feldtkeller equation

$$|H_t(j\omega)|^2 = |K(j\omega)|^2 + 1, \tag{1.31}$$

which in the s domain becomes

$$H_t(s)H_t(-s) = K(s)K(-s) + 1. \tag{1.32}$$

With the use of the polynomials $D(s)$, $P(s)$, and $F(s)$, it is possible to rewrite equation (1.32) in the form

$$D(s)D(-s) = F(s)F(-s) + P(s)P(-s), \tag{1.33}$$

and to calculate, after factorization, the polynomial $F(s)$ on the basis on the given polynomials $N(s)$ and $D(s)$ of the voltage transfer function (1.22). On the basis of the polynomials in the Feldtkeller equation, scattering matrix S_c of a two-port network can be obtained in the form

$$S_c = \frac{1}{D(s)} \begin{bmatrix} F(s) & \pm & P(-s) \\ P(s) & \mp & F(-s) \end{bmatrix}. \tag{1.34}$$

The reflected power P_r can be calculated as

$$P_r = P_{max} - P_1, \tag{1.35}$$

where P_{max} is given by (1.23) and $P_1 = |I_1|^2 reZ_{in}$. Hence

$$P_r = P_{max}\left(1 - \frac{4R_1}{|V_{in}|^2}|I_1|^2 reZ_{in}\right) = P_{max}\left(1 - \frac{4R_1 reZ_{in}}{(R_1 + reZ_{in})^2 + (imZ_{in})^2}\right), \tag{1.36}$$

and after simple calculations we have

$$\frac{P_r}{P_{max}} = \frac{|K(j\omega)|^2}{|H_t(j\omega)|^2} = \left|\frac{R_1 - Z_{in}(j\omega)}{R_1 + Z_{in}(j\omega)}\right|^2. \tag{1.37}$$

The last equation gives the following relation in the s domain

$$\frac{K(s)}{H_t(s)} = \frac{F(s)}{D(s)} = \frac{R_1 - Z_{in}(s)}{R_1 + Z_{in}(s)}, \tag{1.38}$$

or

$$Z_{in}(s) = R_1 \frac{D(s) - F(s)}{D(s) + F(s)}. \tag{1.39}$$

Darlington proved that each positive real function can be realized as a driving-point impedance $Z_{in}(s)$ of a lossless two-port network terminated in a resistor. There are many synthesis methods for the realization of a two-port network with

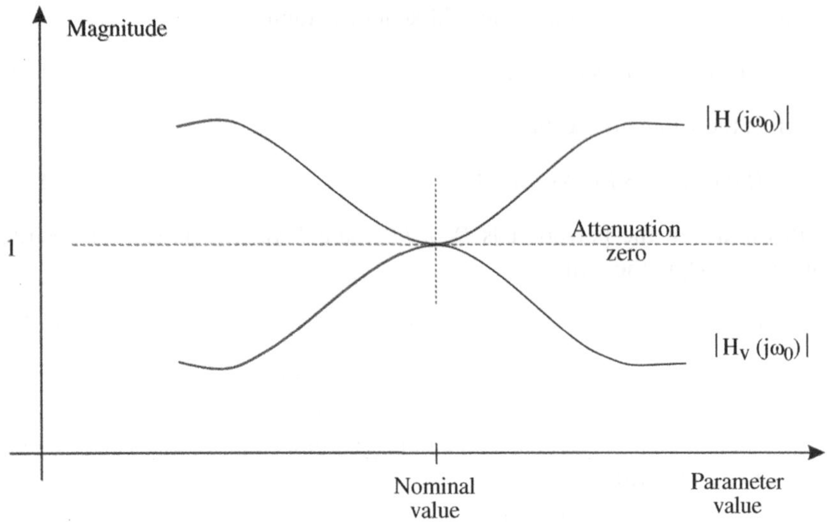

Figure 1.6 Transducer factor and voltage ratio as functions of a lossless circuit parameters.

the use of Z_{in}. In the realization procedure of such a two-port network on the basis of the voltage transfer function (1.22), the Feldtkeller equation plays the main role. Such an approach to synthesis, based on a lossless two-port network, is called the Darlington concept of circuit synthesis.

It follows from relations (1.25) and (1.28) that the magnitude of the transducer factor $H_t(j\omega)$ achieves the minimum value 1 for the frequencies in the passband in which the matching conditions are fulfilled. It means that the attenuation is equal to zero. If we assume $2\sqrt{\frac{G_2}{G_1}} = 1$, the magnitude of the transfer function $H(j\omega)$ achieves the maximum value 1. This case is presented in Fig. 1.6. We see that the derivative of the transfer function magnitude $|H^a|$, and consequently the magnitude sensitivity

$$S_{x_i}^{|H^a|} = \frac{\partial \ln |H^a|}{\partial \ln x_i} = \frac{x_i}{|H^a|} \frac{\partial |H^a|}{\partial x_i}, \tag{1.40}$$

is equal to zero with respect to parameters x_i. The smaller the ripple error in the passband, the smaller the magnitude sensitivity in this band is expected to be.

In Chapters 4 and 5, we will describe synthesis methods of active circuits based on a lossless prototype network. With the use of these methods, we will design active filters that are easy to realize in the OTA-C, the SI, or digital techniques. We assume that such filters are obtained on the basis of the lossless prototype network inherits sensitivity properties of their prototype networks. The most popular prototype circuits are reactance ladder, two-port networks. However, in this design strategy, only the transfer function with a purely even or odd numerator polynomial can be realized (Temes and LaPatra, 1977), as can be observed in Sections 1.2 and 1.3. To synthesize filters and multiport networks like multiplexers, with zeros in the right-hand side of the complex plane, we need nonreciprocal prototype circuits. Hence, the more general gC circuits are considered in this book.

2
Basics of Computer Tools

At the beginning of the chapter, basic information about compiling and running programs in the Linux operating system using a virtual machine is presented. Sections 2.2–2.4 are devoted to a detailed discussion of multidimensional signal processing methods and algorithms. As mentioned in Chapter 1, classes **SNc** and **MDp**, based on these algorithms, will be described.

The most important objects and member functions of the **SNc** class, presented in Section 2.2, are related to the concept of the structural number (SN). This number is represented in the class by a three-dimensional (3D) array *var*[*i*][*j*][*k*]. The elements of this table are related to the symbolic notation of the coefficients of the numerator and denominator polynomials appearing in the transfer functions of the gC circuit with the number of gates *port* and dimension *dimens*. A simple example of such a symbolic notation is the system of equation (1.5).

Section 2.3 presents the mathematical foundations of operations on multidimensional polynomials. Using the method of mathematical induction, a transformation matrix between the analog s and discrete z domains is obtained for multidimensional polynomials. The matrix notation associated with this transformation for one- (1D) and two-dimensional (2D) systems is generalized to the case with any number of dimensions using the tensor notation.

The last section describes the **MDp** class, which builds on the mathematical foundations of Section 2.3. The most important member function of this class is *bili*, which calculates a bilinearly transformed object P for the polynomial and the bilinear transformation array T given as the arguments. Important member functions are also *multiply* and *TimeResp*. The first one is used to design a digital equivalent of the gC circuit working with another digital circuit. The second of the mentioned member functions is needed to calculate the time-domain response of a discrete 1D, 2D, or 3D system to any input signal of the system.

Methods and Tools for Computer-aided Design of Multidimensional Filters in Signal Processing,
First Edition. Andrzej Handkiewicz.

2.1 Basic Operations in Linux and in Matlab or Octave Environments

Most of the tools presented in the monograph are dedicated to work in the Linux system. For a different computer operating system, the first step will be to install a virtual machine, using, e.g. Oracle VirtualBox, and mount a shared directory, e.g. *sharedLin*. Virtual machine names such as *mentor* or *debian*, with the operating system Linux installed, will be used as examples in Sections 2.2–2.4. The path to the shared directory, e.g. $C:\backslash mentor\backslash sharedLin$, must be set beforehand via: *VirtualBox⇒Settings⇒Shared Folders.*

The created shared directory is, after logging in by the administrator with the **su** command, mounted in the following steps:

1. **su**
   enter a password, e.g. debian or centos (the password is invisible when typing),
2. **mount -t vboxsf sharedLin /mnt**,
3. **su debian/centos**.

Using the tools to which this monograph is devoted does not require advanced IT knowledge. Here, we will gather elementary tips for compiling programs on Linux, increasing the stack, deleting a directory, running Octave and Matlab environments:

- compilation in C++: **c++ file.cpp -o outFile**,
- running the compiled program: **./outFile**,
- manual C++ invoke: **man gcc**,
- stack increase: **ulimit -S -s 131072**,
- deleting a directory requires, as in the case of mounting a shared directory, logging in by the administrator in a terminal window (**su**), entering a password, e.g. centos or debian, and then removing the directory with the following command: **rm -r DirectoryName**,
- after installing the Octave environment, download the appropriate packages (e.g. image, signal, etc.), after running the administrator (**su**) in the terminal window:
  su
  apt install octave-image -y
  exit,
- package list can be displayed in the command window with the command: **apt-cache search octave**,
- starting the Octave environment on the *debian* VM: **octave - -gui**, each time you start Octave, type **pkg load video**, **pkg load signal**, etc., if any of these tools to be used,
- starting Matlab on the *debian* VM: type **matlab** in the terminal.

The given names of virtual machines, as mentioned in the introduction, are examples, and the above guidelines are illustrative. When using them, the names of the user's virtual machines and the licenses held should be taken into account. Note that Octave has open access.

2.2 Calculation of Structural Numbers with the Use of Class SNc

For calculation of the object called the SN, there is used a class **SNc**, created in C++, stored in the **SN.h** library available in the directory *src* mentioned in Section 1.1.2. The class is used to detect the following objects in an output file (***.sn**) of the **gCstudio** environment:

port – number of ports of the gC circuit (the transfer functions for these ports are described by SN's),
dimens – dimension of the gC circuit described in $s_1, s_2, \dots, s_{dimens}$ domains,
Bstr – table containing names of gC circuit ports (*Bstr*[*i*], for *i*th SN),
acls – number of the subsets in each SN,
cls – table of line numbers in SN between which the subsets are placed (*cls*[*i*][*j*], for *i*th SN and its *j*th set),
cp – number of columns in SN,
p – table of powers of variables $s_1, s_2, \dots, s_{dimens}$ which correspond to SN subsets (the table is declared as *p*[*port*][*acls*][*dimens*]),
var – 3D array *var*[*i*][*j*][*k*] containing SN elements, where *i* is the port number of gC circuit, *j* is the line number and *k* is the column number in SN. The range of the index *j* is given by *LENTRIES* described in Section 1.1.1, $i = 0$ corresponds to the transfer function denominator and $i = 1, \dots, port - 1$ correspond to the transfer function numerators. The column $k = 0$ contains the coefficients and the columns $k = 1, \dots, cp$ contain the numbers of the parameters *xk* of the circuit in each component of the system of equation (1.5).

The division into sets, corresponding to the exponents of powers of complex variables $s_1, s_2, \dots, s_n$ (*n* – circuit dimension), is shown in the example in Section 1.1.1 with dashed lines.

The listed objects are evaluated as the arguments of the member function *calcSN*. The second basic function of the class is *SNtoPoly*. This member function calculates the coefficients of the transfer function polynomials, associated with the selected port pairs, for the obtained SN and for the given circuit parameter's vector **X**. A class to operate on such multidimensional polynomials will be presented in Section 2.4. The SN calculator **SNc** was created using in particular the **istringstream** class available in the **sstream** library of the C++ language.

The components of the **istringstream** class include, among others, the following functions: **mem** (storing a stream in the cache), **find** (finding a given sequence of characters in the stream, hereinafter referred to as a string, specifying its place in the stream), **length** (calculating the length of a given string), and **seekg** (setting the pointer to a given place in the stream for reading – get). As you can see from the source version of the **SNc** class, these functions are used to look up parameter names and values: *DIMENSION*, *LENTRIES*, *COLUMNS*, *BEGIN*, *END*, *%DRP*, and *%! S* in the cache stream.

2.3 Operations on Multidimensional Polynomials

In this section, we will present formulas for operations on the n-dimensional polynomials, which are functions of operator variables in the analog $s_1, s_2, \dots, s_n$ or discrete $z_1, z_2, \dots, z_n$ domains. We'll start with the 1D case $n = 1$. We will then move on to the 2D case ($n = 2$), which we generalize to the case of any number of dimensions $n > 2$.

The transfer function between two selected terminals of a multiport circuit can be written in matrix notation as

$$H^a(s) = \frac{P'S}{Q'S}, \tag{2.1}$$

where P and Q are matrices of polynomial coefficients. The polynomials are of degree k with respect to the variable s, while S is a vector containing this variable in appropriate powers:

$$S' = [s^0 \;\; s^1 \;\; \dots \;\; s^k]. \tag{2.2}$$

The symbol $'$ means transposition of a vector or matrix. From the transfer function $H^a(s)$, one can find the time-domain impulse response using the inverse Laplace transform. The frequency domain response is possible with $s = j\omega$ as implied by the relationship between the Laplace and Fourier transforms.

As mentioned in the introduction to Chapter 1, the gC circuit can be used for more than just continuous time signal processing. The prototype gC circuits are used to design discrete time counterpart networks implemented in techniques such as the switched current (SI), the switched capacitor (SC), and the digital. Such networks are described in the discrete domain z using the $\mathcal{Z}$ transform. The exact relationship $z = \exp(sT)$ between the transforms $\mathcal{Z}$ and $\mathcal{L}$ (Laplace) replaces, however, a rational function in the domain s by an irrational function in the domain z. The symbol T means the sampling period. We obtain the transfer function as a rational function in the discrete domain z using a bilinear transformation. This transformation approximates the exact relationship $z = \exp(sT)$, resulting from the series expansion of the logarithmic function: $sT = \ln z$. The bilinear transformation has the form:

$$\frac{sT}{2} = \frac{z-1}{z+1} = \frac{1-z^{-1}}{1+z^{-1}}. \tag{2.3}$$

The obtained transfer function will be denoted by $H(z)$, omitting the upper index a in the notation of the transfer function (2.1). Using the bilinear transformation (2.3), we obtain $H(z)$ as a rational function in the domain z, also written in a matrix form:

$$H(z) = \frac{A'Z}{B'Z}, \tag{2.4}$$

where A and B are matrices of polynomial coefficients. The polynomials are of degree k with respect to the variable z^{-1}, while Z is a vector containing this variable in appropriate powers:

$$Z' = [z^0 \;\; z^{-1} \;\; \dots \;\; z^{-k}]. \tag{2.5}$$

The matrices P, Q and A, B are all of size $(k+1) * 1$, that is, they are column vectors with the number of $k+1$ rows. The forms of the vectors S and Z mean that the transfer function polynomials are ordered in ascending order of the powers of the variables s and z^{-1}. Hence, line numbers of the matrices in the range $0, \dots, k$ denote also the powers of the corresponding variables s and z^{-1}, as it is shown in the relations:

$$P'S = \sum_{i=0}^{k} p_i s^i = p_i s^i,$$

$$Q'S = \sum_{i=0}^{k} q_i s^i = q_i s^i,$$

$$A'Z = \sum_{i=0}^{k} a_i z^{-i} = a_i z^{-i},$$

$$B'Z = \sum_{i=0}^{k} b_i z^{-i} = b_i z^{-i}. \tag{2.6}$$

Each of the polynomials in (2.6) is written in two ways: with and without the sum symbol, according to the tensor notation, in which repeated subscripts and superscripts mean automatic summation over their entire range. The superscripts are also the exponents of the powers of s and z^{-1}.

It is shown in Handkiewicz (2002) that the matrices A, B, P, Q are related by the following relationships for 1D networks:

$$P = (A'T_k)', \quad Q = (B'T_k)', \quad A = (P'T_k)', \quad B = (Q'T_k)', \tag{2.7}$$

where T_k is a transformation matrix of size $(k+1) * (k+1)$, realizing a bilinear transformation (2.3). The subscript k in this matrix distinguishes it from the sampling period marked with the letter T without any index. When considering multidimensional systems, we will add a superscript to the sampling period, denoting the selected dimension in which we are sampling.

The matrix T_k can be generated in a recurrent manner:

$$T_0 = [\,1],\quad T_1 = \begin{bmatrix} 1 & 1 \\ 1 & -1 \end{bmatrix},\quad T_2 = \begin{bmatrix} 1 & 2 & 1 \\ 1 & 0 & -1 \\ 1 & -2 & 1 \end{bmatrix},$$

$$T_3 = \begin{bmatrix} 1 & 3 & 3 & 1 \\ 1 & 1 & -1 & -1 \\ 1 & -1 & -1 & 1 \\ 1 & -3 & 3 & -1 \end{bmatrix}, \ldots,\quad T_k. \tag{2.8}$$

The construction procedure of these matrices is as follows. The ith rows, $i = 1, \ldots, k$, of the successive matrices T_j, $j = i-1, \ldots, k$, always form two neighbouring rows of a Pascal triangle with $T_0 = [1]$. For example, the first row of T_0, the first row of T_1, the first row of T_2, the first row of T_3, etc. form the Pascal's triangle. Similarly, elements composed of the second row of T_1, the second row of T_2, the second row of T_3, etc., or the third row of T_2 and the third row of T_3, etc., are also calculated like the Pascal triangle elements. The elements of the last row of each matrix are equal to the elements of the first row with alternately changed signs.

We will prove the relations in (2.8) by mathematical induction. This proof technique needs to show that T_1 is correct and the assumption that T_{k-1} is true implies that T_k is true. Using the bilinear transformation (2.3) for $T = 2$ we can write

$$s = \frac{1-z^{-1}}{1+z^{-1}},\qquad z^{-1} = \frac{1-s}{1+s}. \tag{2.9}$$

Hence, the vector (2.2) can be transformed into discrete domain as

$$\begin{bmatrix} 1 \\ s \\ \vdots \\ s^{k-1} \end{bmatrix} = \frac{1}{(1+z^{-1})^{k-1}} \begin{bmatrix} (1+z^{-1})^{k-1} \\ (1+z^{-1})^{k-2}(1-z^{-1}) \\ \vdots \\ (1-z^{-1})^{k-1} \end{bmatrix}, \tag{2.10}$$

or

$$\begin{bmatrix} 1 \\ s \\ \vdots \\ s^{k-1} \\ s^k \end{bmatrix} = \frac{1}{(1+z^{-1})^{k}} \begin{bmatrix} (1+z^{-1})(1+z^{-1})^{k-1} \\ (1+z^{-1})(1+z^{-1})^{k-2}(1-z^{-1}) \\ \vdots \\ 1+z^{-1})(1-z^{-1})^{k-1} \\ (1-z^{-1})^{k} \end{bmatrix}. \tag{2.11}$$

In each of the matrices on the right side in the above equations, the subsequent elements are polynomials of the variable z^{-1}, with each polynomial in (2.11) equal to the polynomial in (2.10) multiplied by $(1+z^{-1})$. Its coefficients are therefore calculated according to the rule used in Pascal's triangle. The polynomials in the first and last rows of both matrices differ only in sign at the z^{-1} variable, which means a corresponding change in the sign of the coefficient in the last row of the T_k matrix relative to the first row. For $k=1$, these polynomials have coefficients contained in the T_0 and T_1 matrices in (2.8). Assuming, according to mathematical induction, that equation (2.10) written in the form

$$\begin{bmatrix} 1 \\ s \\ \vdots \\ s^{k-1} \end{bmatrix} = \frac{T_{k-1}}{(1+z^{-1})^{k-1}} \begin{bmatrix} 1 \\ z^{-1} \\ \vdots \\ z^{-k+1} \end{bmatrix}, \tag{2.12}$$

holds, we obtain that the equation

$$\begin{bmatrix} 1 \\ s \\ \vdots \\ s^{k-1} \\ s^k \end{bmatrix} = \frac{T_k}{(1+z^{-1})^k} \begin{bmatrix} 1 \\ z^{-1} \\ \vdots \\ z^{-k+1} \\ z^{-k} \end{bmatrix}. \tag{2.13}$$

is also true. This can be written in the symbolic form as

$$S = T_k Z, \tag{2.14}$$

where the scaling factor $1/(1+z^{-1})^k$, which does not affect the transfer function (2.4), has been dropped. We see that both s to z^{-1} and z^{-1} to s transformations in (2.9) have the same form. Hence, similarly to (2.14), we can write

$$Z = T_k S, \tag{2.15}$$

which completes the proof of the relations (2.7).

It is shown in the articles by Handkiewicz et al. (2014) and Bao and Kang (2006) that each transformation matrix can be also directly calculated. The first column elements of the matrix $T_k = [t^i_{k\,j}]$, $i,j = 0,\dots,k$ are all equal to 1 ($t^i_{k\,0} = 1$, $i = 0,\dots,k$) and the first row elements are obtainable from the known, for the Pascal triangle elements, formula

$$t^0_{k\,j} = \binom{k}{j} = \frac{k!}{j!(k-j)!}, \tag{2.16}$$

where $j = 0,\dots,k$. The remaining elements $t^i_{k\,j}$, $i,j = 1,\dots,k$ of the matrix T_k can be obtained from the formula

$$t^i_{k\,j} = t^{i-1}_{k\,j} - (t^{i-1}_{k\,j-1} + t^i_{k\,j-1}), \tag{2.17}$$

where $i, j = 1, \dots, k$. In the above formulas, the upper i and the lower j indices are taken as the row and column numbers of the matrix T_k, respectively.

Example 2.1 *Let us illustrate the relationship (2.17) by calculating the matrix T_2 in (2.8). The first column consists of ones and the first row is calculated from (2.16) as $t^0_{2\ 0} = \binom{2}{0} = 1$, $t^0_{2\ 1} = \binom{2}{1} = 2$, $t^0_{2\ 2} = \binom{2}{2} = 1$. The next element that needs to be calculated is marked in the matrix*

$$T_2 = \begin{bmatrix} 1 & 2 & 1 \\ 1 & ? & \\ 1 & & \end{bmatrix},$$

with the sign ?. We get it from formula (2.17) as $t^1_{2\ 1} = t^0_{2\ 1} - (t^0_{2\ 0} + t^1_{2\ 0}) = 2 - (1 + 1) = 0$. The next two steps is calculation of the elements marked by ? in:

$$T_2 = \begin{bmatrix} 1 & 2 & 1 \\ 1 & 0 & ? \\ 1 & ? & \end{bmatrix}.$$

They are obtained as $t^1_{2\ 2} = t^0_{2\ 2} - (t^0_{2\ 1} + t^1_{2\ 1}) = 1 - (2 + 0) = -1$ and $t^2_{2\ 1} = t^1_{2\ 1} - (t^1_{2\ 0} + t^2_{2\ 0}) = 0 - (1 + 1) = -2$. The last element in

$$T_2 = \begin{bmatrix} 1 & 2 & 1 \\ 1 & 0 & -1 \\ 1 & -2 & ? \end{bmatrix},$$

is calculated as $t^2_{2\ 2} = t^1_{2\ 2} - (t^1_{2\ 1} + t^2_{2\ 1}) = -1 - (0 - 2) = 1$ giving the matrix T_2 in (2.8).

From (2.14) and (2.15), we see that, instead of the inverse matrix T_k^{-1}, the matrix T_k can be used for the inverse z to s transformation. We find that

$$T_k T_k^{-1} = T_k T_k = 2^k U, \tag{2.18}$$

where U is a unit matrix and 2^k is equal to the sum of all binomial coefficients (2.16). Hence, the normalization factor of the matrix T_k is $1/\sqrt{2^k}$.

The representation of the T_k matrix elements, as shown in (2.17), is convenient for the tensor notation of the obtained relations. For example, the first equation in (2.7) can be written as

$$p_j = a_i t^i_{k\ j}, \quad i, j = 0, \dots, k. \tag{2.19}$$

Let us note that the transformation matrix T_k, composed of elements given by (2.8) or (2.16) and (2.17), describes the bilinear transformation (2.3) for the sampling period $T = 2$. Such an assumption simplifies the obtained transformation formulas. However, this scale change in the time domain results in the corresponding change in the frequency ω and $s = j\omega$ domains according to $\omega 2\pi/T$. Hence, before we convert the transfer function from the s into the z^{-1} domain, we ought to scale the transfer function polynomial coefficients by 2^j, where j is an exponent of the variable s of the polynomial. Such rescaling of the coefficients causes that despite changing the frequency scale, the transfer function does not change. For example, let us convert the transfer function $H^a(s) = 1/(1 + s/2 + s^2/4)$ with the coefficients $P' = [1 \ \ 0 \ \ 0]$, $Q' = [1 \ \ 1/2 \ \ 1/4]$ of the numerator and denominator polynomials, respectively, into z^{-1} domain. We substitute $P[j]$ for $2^j P[j]$ ($2^j P[j] \Rightarrow P[j]$), and $Q[j]$ for $2^j Q[j]$ ($2^j Q[j] \Rightarrow Q[j]$), where $j = 0, 1, 2$ to obtain $P' = [1 \ \ 0 \ \ 0]$, $Q' = [1 \ \ 1 \ \ 1]$, and $H^a(s) = 1/(1 + s + s^2)$. In the case of n-dimensional polynomials with coefficients $\alpha_{j_1 \, j_2 \cdots j_n}$, the scaling relationship is in the form of the following substitution:

$$2^{j_1 + j_2 + \cdots + j_n} \alpha_{j_1 \, j_2 \cdots j_n} \Rightarrow \alpha_{j_1 \, j_2 \cdots j_n}. \tag{2.20}$$

We assume that the transfer functions in the following examples are used after such substitution.

Example 2.2 *For both 1st-order and 1st-degree transfer function, we have*

$$H^a(s) = \frac{1}{1+s} = \frac{P'S}{Q'S}, \tag{2.21}$$

where $P' = [1 \ \ 0]$, $Q' = [1 \ \ 1]$, *and* $S' = [1 \ \ s]$, *we have* $A' = P'T_1 = [1 \ \ 1]$, $B' = Q'T_1 = [2 \ \ 0]$. *So we get in the domain* z:

$$H(z) = \frac{A'Z}{B'Z} = \frac{1 + z^{-1}}{2} = \frac{1}{2}(1 + z^{-1}), \tag{2.22}$$

where $Z' = [1 \ \ z^{-1}]$. *Note that the last two equations in (2.7) result in* $(A'T_1)' = 2P$ *and* $(B'T_1)' = 2Q$. *However, the coefficient* 2^1, *resulting from the normalizing relationship (2.18), does not change the exemplary transfer function (2.21).*

Example 2.3 *For the 2nd degree transfer function*

$$H^a(s) = \frac{1}{1 + s + s^2} = \frac{P'S}{Q'S}, \tag{2.23}$$

where $P' = [1 \ \ 0 \ \ 0]$, $Q' = [1 \ \ 1 \ \ 1]$ *and* $S' = [1 \ \ s \ \ s^2]$ *we have* $A' = P'T_2 = [1 \ \ 2 \ \ 1]$, $B' = Q'T_2 = [3 \ \ 0 \ \ 1]$. *Hence, the transfer function is in the* z *domain as follows:*

$$H(z) = \frac{A'Z}{B'Z} = \frac{1 + 2z^{-1} + z^{-2}}{3 + z^{-2}}, \tag{2.24}$$

where $Z' = [1 \ \ z^{-1} \ \ z^{-2}]$. *Similar to the previous example* $(A'T_1)' = 4P$ *and* $(B'T_1)' = 4Q$. *So the scaling factor is* $4 = 2^2$.

Two-dimensional transfer functions can also be presented in the matrix notation. For example, from the relations given in Handkiewicz (2002) and the formula for the transposition of the matrix product $(ab)' = b'a'$ it follows that the transfer function of a 2D system can be written in the form:

$$H^a(s_1, s_2) = \frac{(P'S_1)'S_2}{(Q'S_1)'S_2}, \tag{2.25}$$

where

$$S_i' = [s_i^0 \ s_i^1 \ \dots \ s_i^{k_i}], \quad i = 1, 2, \tag{2.26}$$

and

$$H(z_1, z_2) = \frac{(A'Z_1)'Z_2}{(B'Z_1)'Z_2}, \tag{2.27}$$

where

$$Z_i' = [z_i^0 \ z_i^{-1} \ \dots \ z_i^{-k_i}], \quad i = 1, 2. \tag{2.28}$$

The matrices P, Q, A, and B are all of the order $(k_1 + 1) * (k_2 + 1)$. The relations (2.26) and (2.28) show that the row and column numbers in the range $0, \dots, k_i$ correspond to the powers of the variables s_i and z_i^{-1}, $i = 1, 2$.

With the help of the matrix T_k, you can also perform the bilinear transformation in both dimensions using the formulas

$$P = (A'T_{k_1})'T_{k_2}, \quad Q = (B'T_{k_1})'T_{k_2}, \quad A = (P'T_{k_1})'T_{k_2}, \quad B = (Q'T_{k_1})'T_{k_2}, \tag{2.29}$$

(Handkiewicz, 2002). Each of the above matrix relationships has its equivalent in tensor notation. For example, the third equality in (2.29) can be written as:

$$a_{l\,m} = p_{i\,j} t^{i}_{k_1\,l} t^{j}_{k_2\,m}, \quad i, l = 0 \dots k_1, \quad j, m = 0 \dots k_2. \tag{2.30}$$

Example 2.4 *For the two-dimensional circuit with the 1st-degree transfer function in the both dimensions*

$$H^a(s_1, s_2) = \frac{1}{1 + s_1 + s_2} = \frac{(P'S_1)'S_2}{(Q'S_1)'S_2}, \tag{2.31}$$

where $P = \begin{bmatrix} 1 & 0 \\ 0 & 0 \end{bmatrix}$, $Q = \begin{bmatrix} 1 & 1 \\ 1 & 0 \end{bmatrix}$ *and* $S_1' = [1 \ s_1]$, $S_2' = [1 \ s_2]$ *we have:*

$$A = (P'T_1)'T_1 = \begin{bmatrix} 1 & 0 \\ 1 & 0 \end{bmatrix} T_1 = \begin{bmatrix} 1 & 1 \\ 1 & 1 \end{bmatrix}, \qquad B = (Q'T_1)'T_1 = \begin{bmatrix} 2 & 1 \\ 0 & 1 \end{bmatrix} T_1 = \begin{bmatrix} 3 & 1 \\ 1 & -1 \end{bmatrix}.$$

Hence, we obtain in the discrete domain:

$$H(z_1, z_2) = \frac{(A'Z_1)'Z_2}{(B'Z_1)'Z_2} = \frac{1 + z_1^{-1} + z_2^{-1} + z_1^{-1}z_2^{-1}}{3 + z_1^{-1} + z_2^{-1} - z_1^{-1}z_2^{-1}}, \tag{2.32}$$

where $Z_1' = [1 \ z_1^{-1}]$, $Z_2' = [1 \ z_2^{-1}]$. *Let us note, that similar to the 1D, 2nd-order circuit, we have* $(A'T_1)'T_1 = 4P$, $(B'T_1)'T_1 = 4Q$, *and the scaling factor is* $4 = 2^{1+1}$.

Example 2.5 *For the two-dimensional circuit with the 1st degree in the s_1 domain and 2nd degree in the s_2 domain of the transfer function*

$$H^a(s_1, s_2) = \frac{1}{1 + s_1 + s_2^2} = \frac{(P'S_1)'S_2}{(Q'S_1)'S_2}, \tag{2.33}$$

where $P = \begin{bmatrix} 1 & 0 & 0 \\ 0 & 0 & 0 \end{bmatrix}$, $Q = \begin{bmatrix} 1 & 0 & 1 \\ 1 & 0 & 0 \end{bmatrix}$ *and* $S_1' = [1 \;\; s_1]$, $S_2' = [1 \;\; s_2 \;\; s_2^2]$ *we have:*
$A = (P'T_1)'T_2 = \begin{bmatrix} 1 & 0 & 0 \\ 1 & 0 & 0 \end{bmatrix} T_2 = \begin{bmatrix} 1 & 2 & 1 \\ 1 & 2 & 1 \end{bmatrix}$, $\quad B = (Q'T_1)'T_2 = \begin{bmatrix} 2 & 0 & 1 \\ 0 & 0 & 1 \end{bmatrix} T_2 =$
$\begin{bmatrix} 3 & 2 & 3 \\ 1 & -2 & 1 \end{bmatrix}$. *So we get in the discrete domain:*

$$H(z_1, z_2) = \frac{(A'Z_1)'Z_2}{(B'Z_1)'Z_2} = \frac{1 + z_1^{-1} + 2z_2^{-1} + 2z_1^{-1}z_2^{-1} + z_2^{-2} + z_1^{-1}z_2^{-2}}{3 + z_1^{-1} + 2z_2^{-1} - 2z_1^{-1}z_2^{-1} + 3z_2^{-2} + z_1^{-1}z_2^{-2}}, \tag{2.34}$$

where $Z_1' = [1 \;\; z_1^{-1}]$, $Z_2' = [1 \;\; z_2^{-1} \;\; z_2^{-2}]$. *When calculating the matrices* $(A'T_1)'T_2 = 8P$ *and* $(B'T_1)'T_2 = 8Q$ *there is a scaling factor* $8 = 2^{1+2}$.

The matrix notation is transparent for dependencies concerning circuits with a number of dimensions not greater than $n = 2$. Generalizing the formula (2.30) for any dimension n, we can write the bilinear relationship between the polynomial coefficients in the analog and discrete domains as follows:

$$\delta_{l_1\, l_2 \dots l_n} = \gamma_{i_1\, i_2 \dots i_n} t^{i_1}_{k_1\; l_1} t^{i_2}_{k_2\; l_2} \dots t^{i_n}_{k_n\; l_n}, \quad l_j, i_j = 0 \dots k_j, \quad j = 1 \dots n. \tag{2.35}$$

For $\gamma = a, b, p, q$, we have $\delta = p, q, a, b$, respectively.

Based on the transfer function in the z domain, the response y of the system to any input signal in the discrete time domain can be derived because z_i^{-m} means a delay of m periods in the ith dimension. Obtaining the frequency response is, as indicated by the Fourier transform, possible after making the substitution

$$z_i = exp(j\omega_i T^i), \quad i = 1 \dots n. \tag{2.36}$$

The symbol T^i denotes the sampling period in the ith dimension, unlike the transformation matrix T_k written with a subscript denoting the degree of the transformed polynomial.

2.4 Class MDp of Multidimensional Polynomials

Formulae and algorithms presented in Section 2.3 were implemented in the C++ language in which the class **MDp** (multidimensional polynomials) was developed. The polynomials of the numerators and denominators of the transfer functions related by the bilinear transformation between the domains s and z are computed according to the dependencies (2.19), (2.30), and (2.35). Polynomials are represented in the class by arrays containing their coefficients. Using the class, it is possible to move 1D, 2D, or 3D transfer functions from the s to the z domain (Handkiewicz, 1988, 1991; Handkiewicz et al., 2014).

The class operates on the following objects:

n – order of the n-dimensional polynomial (note that the class is implemented for $n \leq 3$),

k – degrees of the polynomials in each dimension, defined as elements of the array $k[n]$,

P – array of coefficients, calculated for the current values of the circuit parameters, of the polynomial corresponding to the selected SN; consecutive indices of the array $P[k[1]] \ldots [k[n]]$ are equal to the powers of the variables s or z^{-1}, as marked in (2.26) and (2.28),

y – sample responses for transfer functions formed from polynomials corresponding to their numerators and the denominator (y is a 3D array, because the class is built for gC circuits with dimension no greater than 3),

T – array of bilinear transformation whose elements are described by formulae (2.16) and (2.17), (array $T[m][i][j]$, $m = 1, \ldots, n$, $i, j = 0, \ldots, k[m]$).

The following member functions are available in the class:

readPoly – the function reads the polynomial coefficients from the file of the name given as the function argument,

writePoly – the function writes the polynomial coefficients to the file of the name given as the function argument,

move – the function moves the polynomial given as the argument into the $P[k[1]] \ldots [k[n]]$ object,

T2to1 – the function compensates the $T = 2$ value which is used in the bilinear transformation (2.3),

reverse – the function reverses, with respect to powers of s or z^{-1} variables, the coefficients of the polynomial being an argument of the function and saves as a new object P,

biliT – the function calculates the array T for bilinear transformation given by formulae (2.16) and (2.17), ($arrayT[m][i][j], m = 1 \ldots n, i, j, = 0 \ldots k[m]$),

bili – the function calculates a bilinearly transformed object P for the polynomial and the bilinear transformation array T given as the arguments,

multiply – the function multiplies two polynomials given as the arguments of the function,

TimeResp – the function calculates samples of a three-dimensional response $y[k1][k2][k3]$, $k1 = 0, \ldots, N[0] - 1$, $k2 = 0, \ldots, N[1] - 1$, $k3 = 0, \ldots, N[2] - 1$, where $N[0]$, $N[1]$, $N[2]$ are a number of samples in each dimension. The samples of the input signal, the numerator, and the denominator polynomials of the transfer function are given as the arguments of *TimeResp*. For the case of one-dimensional, we simply substitute $N[1] = N[2] = 1$ and $N[0]$ equal to the number of samples in that dimension.

3

Integrated Circuit Design

The chapter begins with a brief presentation of two automatic integrated circuit design environments, Cadence and Mentor Graphics. The chapter, for the most part, is devoted to the switched current (SI) technique, which appears to be most suitable for implementing analog circuits in standard digital nanometer-scale technologies. We will begin this part with a description of another environment **SIstudio**, which allows automatic generation of the SI counterpart of the prototype gC circuit. Next, the basic properties of SI circuits will be discussed. All these circuits are based on complementary transistor pairs, which can be stacked in currently developed nanometer technologies (three-dimensional [3D] stacked CMOS). The n-type transistor is, in the device layer processing, built above the p-type transistor (Huang et al., 2020).

The working principle of basic cells, the complementary transistor pair and the current mirror, are discussed based on the square-law FET model. These circuits are, in turn, used to build two types of current-mode continuous-time and SI integrators. Sections 3.6 and 3.7 examine the basic parameters of these macrocells: thermal noise, distortion, common mode gain, and common mode rejection ratio (CMRR). Sections 3.8 and 3.9 are devoted to the differential stage and the transconductance amplifier (OTA), enabling direct implementation of gC circuits.

3.1 Design Automation of Integrated Circuit

Cadence and Mentor Graphics are among the most widely used environments for automatic IC layout generation. Mentor Graphics is a licensed tool that enables full automation of designing digital integrated circuits. The starting point is the circuit shown on the diagram using the appropriate graphical interface or described in text, e.g. in VHDL. Based on technology files provided by IC manufacturers, tools like Mentor Graphics enable the automatic generation of a digital IC based on a given description. One way to represent circuit topography is in Mentor Graphics'

Methods and Tools for Computer-aided Design of Multidimensional Filters in Signal Processing,
First Edition. Andrzej Handkiewicz.

AMPLE package. In contrast to the digital circuit, however, obtaining the layout of the analog circuit requires a large, direct contribution of the designer's work. There is a lack of universal design tools for analog circuits as it is for digital circuits.

As mentioned in the introduction to Chapter 1, one possible implementation of the prototype gC circuits is the use of the SI technique. An example of using this technique will be presented in Chapter 4. The **SIstudio** environment developed for this purpose allows, based on the description in VHDLA, to generate a circuit with SI, as described in Szczesny et al. (2012). The layout generation presented in this chapter was possible thanks to the description of the SI circuit in the AMPLE environment. Full automation of the layout design of the analog SI circuit, obtained in this way, was achieved using the **ICstudio** tool. It is available in the Mentor Graphics environment, enabling the generation of a layout described in the AMPLE language. A detailed description of **SIstudio** can be found in Section 4.4, where the example of the analysis filter shows its implementation in the technique of SI.

3.2 Current Mode Integrated Circuit Design

As mentioned, technology development is currently oriented to digital circuits. It is almost impossible to realize basic classical analog cells like op-amps and various kinds of amplifiers in nanoscale low-voltage supply technology. The analog circuits that can operate on the same chip with digital part ought to have the following properties:

1. operation at low-voltage supply,
2. low power consumption,
3. low noise,
4. low sensitivity to technology parameter variations,
5. reduced parasitic influence, and
6. good properties with respect to intermodulation and harmonic distortion

(Sanchez-Sinencio and Andreou, 1999).

We will show that the basic cell that is able to meet such requirements is a complementary pair of n-channel and p-channel transistors in two kinds: diode and inverter connections. It is proved that such circuits are very convenient with respect to nonlinear distortions. A current mirror composed of such circuits is the cell which, for carefully chosen dimensions of transistor channels, precisely operates at low voltage supply. In such approach, the bias current is not directly controlled. Hence, to achieve low power consumption and low noise at current mode operation, the transistor channels have to be as narrow as possible in the chosen technology. Memory cells, continuous and discrete time current mode integrators obtained in this design strategy inherit the above properties from current mirrors. The fully differential structure of circuits composed of such cells allows to reduce parasitic influence. The circuits designed as current mode counterparts of lossless prototype circuits are low sensitive with respect to technology parameter variations.

3.3 Square-Law Model of Field Effect Transistors

Symbols of two kinds of field effect transistors (FETs), composed of a p-channel or an n-channel, are presented on the top or the bottom of Fig. 3.1, respectively. In the most recent technology, the channels are built from nano-sheets (NS) stacked horizontally between source (S) and drain (D) areas (Johnson, 2021). This kind of a FET is called NSFET. Very short channels, with the minimum lengths of 2 nm, are achieved in this technology. The sheets are of 5 nm thickness and from 15 to 70 nm width. The sheets are all around by gates, which are voltage supplied and isolated from the channels by the oxide. In Fig. 3.1, the drains of both transistors are connected together in the node with the potential V. The sources are connected to supply lines V_{DD} and V_{SS} for the p-channel and n-channel NSFETs, respectively. A drain current I_D is controlled by voltages V_{DS} and V_{GS}, and the transistor can be in a cutoff mode ($I_D = 0$) or in an active mode ($I_D \neq 0$), depending on the parameter called the threshold voltage, denoted by V_{Tn} or V_{Tp} for the corresponding channel kind. For example, the n-channel NSFET is *off* (cutoff) when $V_{GS} \leq V_{Tn}$, and it is *on* (active) when $V_{GS} > V_{Tn}$. We can say that in the active mode, the appropriate gate potential induces the conducting channel between the drain and the source.

The simplest mathematical model of FET transistors in the active mode is the square-low one. For an n-channel NSFET, it is assumed in this model, that the drain current depends on drain-source voltage V_{DS} in the first and second power,

$$I_{Dn} = \frac{\beta_n}{2}[2(V_{GS} - V_{Tn})V_{DS} - V_{DS}^2], \tag{3.1}$$

exclusively. It is obvious that the device transconductance β_n is proportional to the width W and inversely proportional to the length L of the transistor channel and can be expressed as

$$\beta_n = k_n' \frac{W}{L} = \mu_n C_{ox} \frac{W}{L}, \tag{3.2}$$

where k_n' is described by physical parameters:

μ_n – the electron mobility (μ_p – hole mobility for p-channel transistors),
C_{ox} – the oxide capacitance per unit gate area.

The width of the channel depends on number of nano-sheets, ns, and can be calculated as $W = ns(2t + 2w)$, where t is the thickness and w is the width of the sheet. The unit area oxide capacitance can be calculated from

$$C_{ox} = \frac{\epsilon_{ox}}{x_{ox}}, \tag{3.3}$$

where $\epsilon_{ox} = 3.9\epsilon_0$, $\epsilon_0 = (8.854E - 14)$ F/m, and x_{ox} is the gate oxide thickness.

The coefficient of the first component in formula (3.1) is linearly dependent on the gate-to-source voltage V_{GS}. For a given voltage V_{GS}, equation (3.1) describes a parabola. Only the part of the parabola between the origin of the coordinate system (I_{Dn} versus V_{DS}) and the peak current of the parabola is used as the current to voltage characteristic. We say that when the values of current and voltages are in this region, the transistor is in a nonsaturated mode. It is easy to calculate from the extreme condition, $\partial I_{Dn}(V_{DS})/\partial V_{DS} = 0$ that the saturation voltage has the value

$$V_{DS,sat} = V_{GS} - V_{Tn}, \tag{3.4}$$

for which the value of the current is

$$I_{Dn} = \frac{\beta_n}{2}(V_{GS} - V_{Tn})^2. \tag{3.5}$$

A more exact relation for the current in the saturated mode contains an additional component

$$I_{Dn} = \frac{\beta_n}{2}(V_{GS} - V_{Tn})^2[1 + \lambda_n(V_{DS} - V_{DS,sat})], \tag{3.6}$$

with λ_n being the n-channel-length modulation parameter.

The equation of a p-channel NSFET in the nonsaturated mode can be written in the complementary to (3.1) form:

$$I_{Dp} = \frac{\beta_p}{2}[2(V_{SG} + V_{Tp})V_{SD} - V_{SD}^2], \tag{3.7}$$

where the current is oriented as shown in Fig. 3.1. The threshold voltage V_{Tp} now has a negative value, and $V_{GS} = -V_{SG} < V_{Tp}$. Similarly, in the saturated mode we have:

$$I_{Dp} = \frac{\beta_p}{2}(V_{SG} + V_{Tp})^2, \tag{3.8}$$

or

$$I_{Dp} = \frac{\beta_p}{2}(V_{SG} + V_{Tn})^2[1 + \lambda_p(V_{SD} - V_{SD,sat})], \tag{3.9}$$

where

$$V_{SD,sat} = V_{SG} + V_{Tp}. \tag{3.10}$$

3.4 Complementary Transistor Pair

Current mode circuits, which we consider, are composed of the complementary pair of transistors presented in Fig. 3.1, as basic cells. We assume that both transistors work in saturation regions

$$V_{DS} = V - V_{SS} \geq V_{DS,sat} = V_{GS} - V_{Tn} = V_{gn} - V_{SS} - V_{Tn}, \tag{3.11}$$

$$V_{SD} = V_{DD} - V \geq V_{SD,sat} = V_{SG} + V_{Tp} = V_{DD} - V_{gp} + V_{Tp}. \tag{3.12}$$

Let us note that for a diode connection (gates and drains of FET transistors connected together) the relations (3.11) and (3.12) are always fulfilled.

For the saturated transistors in Fig. 3.1, the drain currents can be described with the use of a simple quadratic model (3.5), (3.8), as:

$$I_{Dn} = \frac{\beta_n}{2}(V_{GS} - V_{Tn})^2 = \frac{\beta_n}{2}(V_{gn} - V_{SS} - V_{Tn})^2, \tag{3.13}$$

$$I_{Dp} = \frac{\beta_p}{2}(V_{SG} + V_{Tp})^2 = \frac{\beta_p}{2}(V_{DD} - V_{gp} + V_{Tp})^2. \tag{3.14}$$

Assuming

$$V_{DD} = -V_{SS} = V_s, \qquad V_{gp} = V_{gn} = V_g, \tag{3.15}$$

we obtain

$$\begin{aligned} I = I_{Dn} - I_{Dp} &= \frac{1}{2}(\beta_n - \beta_p)V_g^2 \\ &+ [\beta_n(V_s - V_{Tn}) + \beta_p(V_s + V_{Tp})]V_g \\ &+ \frac{1}{2}[\beta_n(V_s - V_{Tn})^2 - \beta_p(V_s + V_{Tp})^2]. \end{aligned} \tag{3.16}$$

Figure 3.1 Complementary pair of n-channel and p-channel transistors.

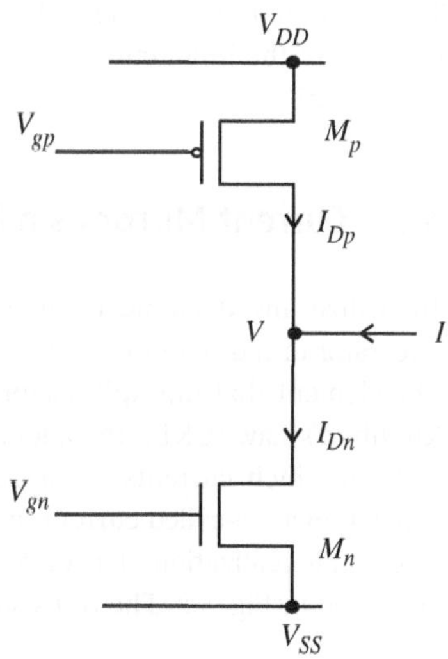

The first component in equation (3.16) describes the distortion of the complementary pair. The second component describes the linear dependence of the current I with respect to the voltage V_g, where the transconductance is given by $[\beta_n(V_s - V_{Tn}) + \beta_p(V_s + V_{Tp})]$. The third component denotes the dc offset of the cell. The dc offset can be effectively canceled when circuits are realized in a balanced structure. The distortion can be canceled for $\beta_n = \beta_p$ or $k'_n W_n/L_n = k'_p W_p/L_p$. Let us note that for $\beta_n = \beta_p$, nearly the same levels of the voltages at the input and the output of the complementary pair can be achieved. It means that no lever shifters are necessary to minimize transmission errors between cells in cascade connection.

Assumptions in (3.15) correspond to the connection of both complementary transistor gates to the common input source. For the circuit in which the gate of the p-channel transistor is connected to the bias voltage V_B, adequate assumptions are

$$V_{DD} = -V_{SS} = V_s, \quad V_{gp} = V_B, \quad V_{gn} = V_g, \tag{3.17}$$

for which we obtain:

$$\begin{aligned} I = I_{Dn} - I_{Dp} &= \frac{\beta_n}{2} V_g^2 \\ &+ \beta_n(V_s - V_{Tn})V_g \\ &- \frac{\beta_p}{2} V_B^2 + \beta_p(V_s + V_{Tp})V_B \\ &+ \frac{\beta_n}{2}(V_s - V_{Tn})^2 - \frac{\beta_p}{2}(V_s + V_{Tp})^2. \end{aligned} \tag{3.18}$$

The comparison of equations (3.16) and (3.18) shows that in the second case the distortion is greater, and the transconductance is smaller than in the first case. The dynamic range of the cell in the first case is twice as great because of the partition of the input current I into two branches, increasing the signal-to-noise ratio by $6dB$.

3.5 Current Mirror as a Parameterized Cell

To realize any dynamic linear system there are necessary, in addition to an integrator or a differentiator, two next elements: an adder and a scaling element (the element that multiplies signals by a constant value). According to Current Kirchhoff's Law (CKL), the adder in a current mode network is simple a circuit node in which currents are added. Also, the scaling element has very simple realization as so-called current mirror.

A simple realization of a current mirror composed of two n-channel NSFETs is presented in Fig. 3.2. The transistor $M0$ is in diode connection, whereas the gate

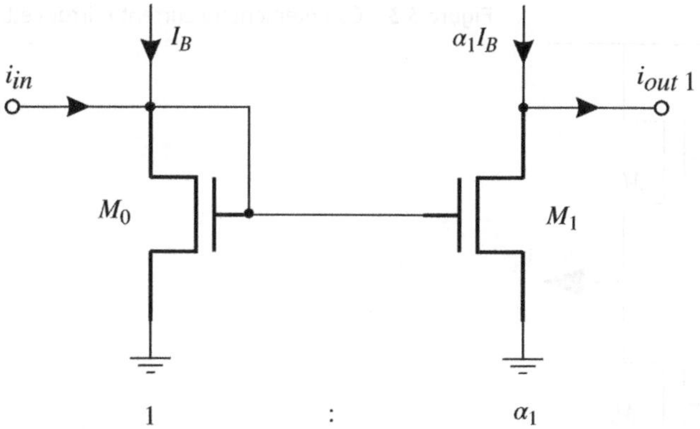

Figure 3.2 Current mirror.

of the transistor $M1$ is connected with the gate of $M0$ and $V_{GS0} = V_{GS1} = V_{GS}$. The coefficient α_1 is determined by the aspect ratios of both transistors

$$\alpha_1 = \frac{W_1/L_1}{W_0/L_0}. \tag{3.19}$$

Assuming that the bias current I_B keeps both transistors in a saturated mode, we have

$$\frac{\alpha_1 I_B - i_{out1}}{I_B + i_{in}} = \frac{\beta_1(V_{GS} - V_{Tn})^2}{\beta_0(V_{GS} - V_{Tn})^2}, \tag{3.20}$$

or

$$\frac{\alpha_1 I_B - i_{out1}}{I_B + i_{in}} = \frac{W_1/L_1}{W_0/L_0} = \alpha_1, \tag{3.21}$$

and

$$i_{out1} = -\alpha_1 i_{in}. \tag{3.22}$$

If $\alpha_1 = 1$, then the circuit mirrors the input current at its output.

Current mirrors with multiple outputs can be obtained by adding transistors $M2 \ldots Mn$ connected to $M0$ in the same manner as $M1$ and described by the coefficients

$$\alpha_i = \frac{W_i/L_i}{W_0/L_0}, \qquad i = 2, \ldots, n. \tag{3.23}$$

In this case, we have

$$i_{outk} = -\alpha_k i_{in}, \qquad k = 2, \ldots, n. \tag{3.24}$$

This equation shows that the current mirrors can be used as multioutput parameterized cells, which are necessary to realize coefficients in linear equations.

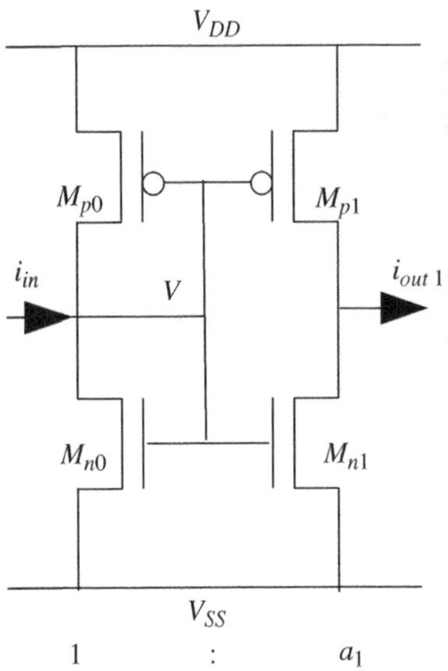

Figure 3.3 Complementary current mirror cell.

A complementary current mirror cell can be obtained with the use of a p-channel NSFETs. A comparison of relations (3.16) and (3.18) shows that to realize the current mirror cell with better behavior with respect to distortion, both the n-channel and p-channel type current mirrors ought to be connected into complementary pairs, as it is shown in Fig. 3.3. Such current mirror cells will be used to build current mode integrators presented in Sections 3.6 and 3.7.

3.6 Continuous-Time Current Mode Integrator

The structure of the continuous-time current mode integrator is presented in Fig. 3.4. The integrator possesses two differential inputs and outputs, hence, it is called a fully differential integrator. As it is easy to recognize, the integrator is in both signal paths and is composed of two output complementary current mirrors.

The drain currents in the saturation mode of n-channel and p-channel transistors can be described by (3.5) and (3.8), where $V_{GS} = V_{1,2} - V_{SS}$, $V_{SG} = V_{DD} - V_{1,2}$. We assume, without loss of generality and for simplification of calculations, that the supply voltages are symmetrical, $V_{DD} = -V_{SS} = V_s$, as well as the gate voltages $V_1 = -V_2 = V_g$. Hence, the drain currents have the form

$$I_{Dn} = \frac{\beta_n}{2}(\pm V_g + V_s - V_{Tn})^2, \qquad I_{Dp} = \frac{\beta_p}{2}(V_s \mp V_g + V_{Tp})^2, \tag{3.25}$$

where the upper signs are taken for transistors in the positive (upper) signal path and the lower signs for transistors in the negative (lower) signal path of the circuit

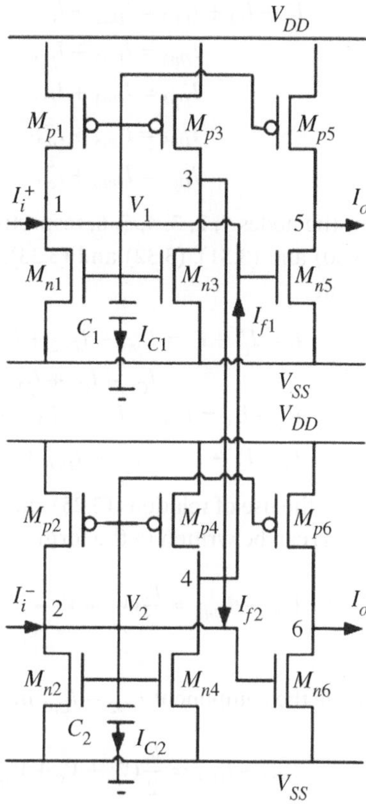

Figure 3.4 Continuous-time current mode integrator.

in Fig. 3.4. If supply is asymmetrical, $V_{DD} \neq |V_{SS}|$, then all voltages in the fully differential circuit are shifted by $V_s = (V_{DD} - V_{SS})/2$.

All transistor pairs used in this circuit have properly sized channel dimensions so that $\beta_n = \beta_p$. For the fully differential structure, we can assume all circuit elements in both positive and negative paths are the same. Hence,

$$\beta_{n1} = \beta_{p1} = \beta_{n2} = \beta_{p2} = \beta_i,$$
$$\beta_{n3} = \beta_{p3} = \beta_{n4} = \beta_{p4} = \beta_f,$$
$$\beta_{n5} = \beta_{p5} = \beta_{n6} = \beta_{p6} = \beta_o. \tag{3.26}$$

Similarly, the integrating capacitances are

$$C_1 = C_2 = C. \tag{3.27}$$

For the circuit in Fig. 3.4, we can write the equations obtained with the use of CKL in the form

$$I_i^+ + I_{f1} + I_{Dp1} = I_{Dn1} + I_{C1}, \tag{3.28}$$

$$I_i^- + I_{f2} + I_{Dp2} = I_{Dn2} + I_{C2}, \tag{3.29}$$

$$I_{Dp3} = I_{Dn3} + I_{f2}, \tag{3.30}$$

$$I_{Dp4} = I_{Dn4} + I_{f1}, \tag{3.31}$$

$$I_{Dp5} = I_{Dn5} + I_o^+, \tag{3.32}$$

$$I_{Dp6} = I_{Dn6} + I_o^-, \tag{3.33}$$

for the nodes 1, 2, 3, 4, 5, 6, respectively. The pairs of equations (3.28) and (3.29), (3.30) and (3.31), (3.32) and (3.33), after the subtraction of both sides, give the result

$$I_i = I_i^+ - I_i^- = I_{Dn1} - I_{Dn2} + I_{Dp2} - I_{Dp1} + I_{C1} - I_{C2} + I_{f2} - I_{f1}, \tag{3.34}$$

$$I_{f2} - I_{f1} = I_{Dp3} - I_{Dp4} + I_{Dn4} - I_{Dn3}, \tag{3.35}$$

$$I_o = I_o^+ - I_o^- = I_{Dp5} - I_{Dp6} + I_{Dn6} - I_{Dn5}. \tag{3.36}$$

With the use of relations (3.25) and (3.26), the component $I_{Dn1} - I_{Dn2}$ in equation (3.34) can be written in the form

$$I_{Dn1} - I_{Dn2} = \frac{\beta_i}{2}(V_g + V_s - V_{Tn})^2 - \frac{\beta_i}{2}(-V_g + V_s - V_{Tn})^2 = 2\beta_i(V_s - V_{Tn})V_g, \tag{3.37}$$

while the component $I_{Dp2} - I_{Dp1}$ in the form

$$I_{Dp2} - I_{Dp1} = \frac{\beta_i}{2}(V_s + V_g + V_{Tp})^2 - \frac{\beta_i}{2}(V_s - V_g + V_{Tp})^2 = 2\beta_i(V_s + V_{Tp})V_g. \tag{3.38}$$

Similarly, we can obtain the components

$$I_{Dp3} - I_{Dp4} = -2\beta_f(V_s + V_{Tp})V_g, \tag{3.39}$$

$$I_{Dn4} - I_{Dn3} = -2\beta_f(V_s - V_{Tn})V_g, \tag{3.40}$$

$$I_{Dp5} - I_{Dp6} = -2\beta_o(V_s + V_{Tp})V_g, \tag{3.41}$$

$$I_{Dn6} - I_{Dn5} = -2\beta_o(V_s - V_{Tn})V_g, \tag{3.42}$$

in remaining equations (3.35) and (3.36). Hence, equations (3.34), (3.35), and (3.36) can be written as

$$I_i = 2\beta_i(2V_s - V_{Tn} + V_{Tp})V_g + 2sCV_g + I_{f2} - I_{f1}, \tag{3.43}$$

$$I_{f2} - I_{f1} = -2\beta_f(2V_s - V_{Tn} + V_{Tp})V_g, \tag{3.44}$$

$$I_o = -2\beta_o(2V_s - V_{Tn} + V_{Tp})V_g, \tag{3.45}$$

if the currents I_{C1} and I_{C2} are changed by $I_{C1} = sCV_1 = sCV_g$ and $I_{C2} = sCV_2 = -sCV_g$ for $C_1 = C_2 = C$. After simple algebraic calculation, we obtain the transfer function of the circuit in Fig. 3.4 in the form

$$H(s) = \frac{I_o(s)}{I_i(s)} = -\frac{\beta_o(2V_s - V_{Tn} + V_{Tp})}{sC + (\beta_i - \beta_f)(2V_s - V_{Tn} + V_{Tp})}. \tag{3.46}$$

This transfer function can be written as

$$H(s) = -\frac{1}{s\tau + (a_i - a_f)}, \tag{3.47}$$

where

$$\tau = \frac{C}{\beta_o(2V_s - V_{Tn} + V_{Tp})}, \tag{3.48}$$

denotes the time constant and

$$a_i = \frac{\beta_i}{\beta_o}, \quad a_f = \frac{\beta_f}{\beta_o}, \tag{3.49}$$

denote the attenuations of the input and feedback stages.

3.6.1 Thermal Noise and Distortions

Comparison of the relations (3.16) and (3.18) shows that the integrator with both types n-channel and p-channel transistors used to operate on current signals have better properties with respect to intermodulation and harmonic distortions with respect to the integrator in which the p-channel transistor is used only to generate a bias current. According to (3.16), the quadratic component can be cancelled for $\beta_n = \beta_p$. Also, the dynamic range of the modified integrator is twice as great, because of the partition of the input currents I_i^+, I_i^- into two branches, increasing the signal-to-noise ratio by 6 dB. The spectrum of the thermal voltage noise generator, referred to as the input of an FET transistor, is given by

$$S_{v_{n,th}^2} = \frac{8KT}{3g_m}\Delta f, \tag{3.50}$$

(Maloberti, 2001), where g_m can be obtained from (3.13) as

$$g_m = dI_{Dn}/dV_{GS} = \beta_n(V_{GS} - V_{Tn}) = k_n' W_n/L_n(V_{GS} - V_{Tn}). \tag{3.51}$$

However, for current mode circuits, we have the spectrum of the current noise generator described by

$$S_{i_{n,th}^2} = S_{v_{n,th}^2} g_m^2 = \frac{8KTg_m}{3}\Delta f. \tag{3.52}$$

We see that this spectrum is in current mode circuits, as opposed to voltage mode circuits, directly proportional to the width and inversely proportional to the length of a transistor channel. Hence, our design strategy is based on transistors with constricted and lengthened channels, and the thermal noise can be also reduced.

3.6.2 Common Mode Gain Analysis

The pairs of equations (3.28) and (3.29), (3.30) and (3.31), (3.32) and (3.33), after the addition of their both sides, give the results

$$I_{i,com} = I_i^+ + I_i^- = I_{Dn1} + I_{Dn2} - (I_{Dp1} + I_{Dp2}) - (I_{f1} + I_{f2}), \tag{3.53}$$

$$I_{f1} + I_{f2} = I_{Dp3} + I_{Dp4} - (I_{Dn3} + I_{Dn4}), \tag{3.54}$$

$$I_{o,com} = I_o^+ + I_o^- = I_{Dp5} + I_{Dp6} - (I_{Dn5} + I_{Dn6}), \tag{3.55}$$

where $I_{C1} + I_{C2} = sCV_g - sCV_g = 0$ for $C_1 = C_2 = C$.

With the use of relations (3.25) and (3.26), the component $I_{Dn1} + I_{Dn2}$ in equation (3.53) can be written in the form

$$\begin{aligned} I_{Dn1} + I_{Dn2} &= \frac{\beta_i}{2}(V_g + V_s - V_{Tn})^2 + \frac{\beta_i}{2}(-V_g + V_s - V_{Tn})^2 \\ &= \beta_i[V_g^2 + (V_s - V_{Tn})^2], \end{aligned} \tag{3.56}$$

while the component $I_{Dp1} + I_{Dp2}$ in the form

$$\begin{aligned} I_{Dp1} + I_{Dp2} &= \frac{\beta_i}{2}(V_s - V_g + V_{Tp})^2 + \frac{\beta_i}{2}(V_s + V_g + V_{Tp})^2 \\ &= \beta_i[V_g^2 + (V_s + V_{Tp})^2]. \end{aligned} \tag{3.57}$$

Similarly, we can obtain the components

$$I_{Dp3} + I_{Dp4} = \beta_f[V_g^2 + (V_s + V_{Tp})^2], \tag{3.58}$$

$$I_{Dn3} + I_{Dn4} = \beta_f[V_g^2 + (V_s - V_{Tn})^2], \tag{3.59}$$

$$I_{Dp5} + I_{Dp6} = \beta_o[V_g^2 + (V_s + V_{Tp})^2], \tag{3.60}$$

$$I_{Dn5} + I_{Dn6} = \beta_o[V_g^2 + (V_s - V_{Tn})^2], \tag{3.61}$$

in remaining equations (3.54) and (3.55). Hence, equations (3.53), (3.54), and (3.55) can be written as

$$I_{i,com} = \beta_i[(V_s - V_{Tn})^2 - (V_s + V_{Tp})^2] - (I_{f1} + I_{f2}), \tag{3.62}$$

$$I_{f1} + I_{f2} = \beta_f[(V_s + V_{Tp})^2 - (V_s - V_{Tn})^2], \tag{3.63}$$

$$I_{o,com} = \beta_o[(V_s + V_{Tp})^2 - (V_s - V_{Tn})^2], \tag{3.64}$$

or

$$I_{i,com} = (\beta_i + \beta_f)(2V_s - V_{Tn} + V_{Tp})(-V_{Tn} - V_{Tp}), \tag{3.65}$$

$$I_{o,com} = -\beta_o(2V_s - V_{Tn} + V_{Tp})(-V_{Tn} - V_{Tp}). \tag{3.66}$$

After simple algebraic calculation, we obtain the common mode gain of the circuit in Fig. 3.4 in the form

$$H_{com} = \frac{I_{o,com}}{I_{i,com}} = -\frac{\beta_o}{\beta_i + \beta_f} = -\frac{1}{a_i + a_f}, \tag{3.67}$$

where the attenuations a_i and a_f of the input and feedback stages are described by (3.49).

3.6.3 Common Mode Rejection Ratio

The CMRR is usually defined using a ratio of the transfer function to the common mode gain of the circuit. Hence, on the basis of formulas (3.47) and (3.67), we have

$$CMRR = 20\lg\frac{|H(j\omega)|}{|H_{com}(j\omega)|} = 20\lg\frac{a_i + a_f}{|\omega|\tau}. \tag{3.68}$$

The CMRR is equal to zero ($CMRR = 0$) for $\omega_{com,0} = (a_i + a_f)/\tau = (a_i + a_f)\omega_1$, where $\omega_1 = 1/\tau$ is the unity gain frequency of the integrator. The enlarging of the attenuations a_i and a_f increases $\omega_{com,0}$. However, the bigger the attenuations are, the smallest are the output currents and the real circuit can be unstable. Hence, an additional circuit which improves CMRR can be used (Szczesny, 2017).

3.7 Switched-Current Integrator

The balanced structure of the bilinear SI integrator is shown in Fig. 3.5 (Hughes et al., 1996). Each pair of blocks (N1, P1 and N2, P2) is composed of complementary NSFETs and gate switches, which are driven by the clock signals Φ_1 and Φ_2, respectively. The connection of complementary transistors and gate switches depends on the kind of second-generation SI memory cell, used as the basis of the integrator, presented in Fig. 3.6. Let us note that each of the memory cell is realization of the current mirror in Fig. 3.2. In the phases $\Phi_1 = 1$ and $\Phi_2 = 0$, the memory cell realises the input stage and in the phases $\Phi_1 = 0$ and $\Phi_2 = 1$ the output stage of the memory cell.

To analyze the balanced integrator, we will first consider the circuit presented in Fig. 3.7, based on the memory cell from Fig. 3.6c. As was mentioned, this memory cell is realization of complementary current mirrors, hence, the output current of the memory cell is a delayed version of the input current, as it is described by equation (3.22). The ratios $1 : 1 : 1 : 1$ at the bottom of the figure refer to the dimensions of the n-channel and p-channel transistors, respectively. As it was in the case of the continuous-time current mode integrator described in Section 3.6, we assume that the transistors operate in saturation regions. We also assume that

$$i_{in+}(n) = -i_{in-}(n) = i_{in}(n), \tag{3.69}$$

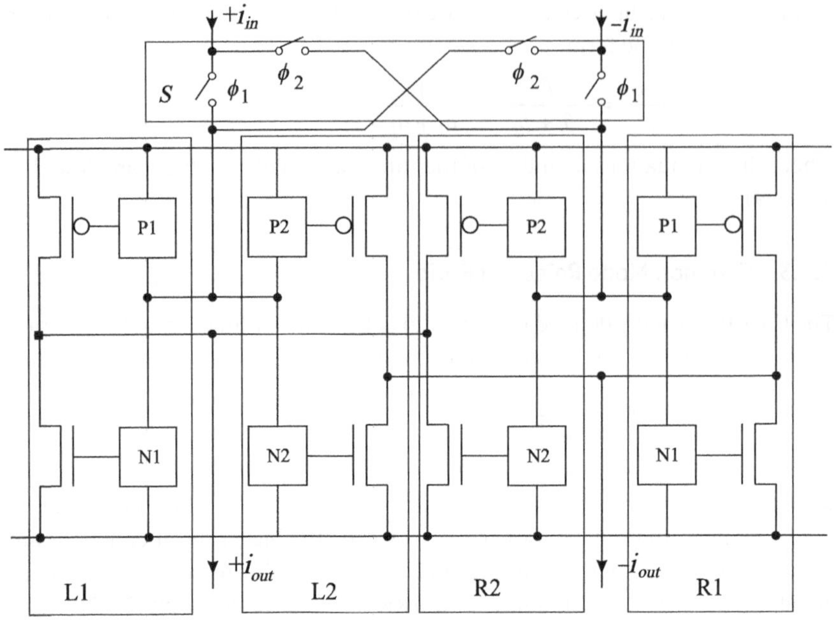

Figure 3.5 Balanced structure of the bilinear integrator.

and that logic "1" of Φ_2 corresponds to the $n-1$ sampling period, while logic "1" of Φ_1 corresponds to the nth sampling period of the input and output signals. Let us note that the sampling frequency is two times greater than the clock frequency.

Let us consider phase Φ_2, corresponding to the $n-1$ sampling period, in which the switches driven by Φ_2 are closed, and those controlled by Φ_1 are open. In this phase, the transistors M_{n2}, M_{p2}, M_{n4}, and M_{p4} compose the current mirror so that the output current can be calculated as:

$$i_{out-}(n) = -[i_{in-}(n-1) + i_{Dp1} - i_{Dn1}]. \tag{3.70}$$

The output current i_{out-} holds the same value after opening the switches controlled by Φ_2. Hence, the relation $i_{out-}(n) = i_{out-}(n-1)$ is used in the above equation. The output current i_{out+} can be expressed as

$$i_{out+}(n-1) = i_{D3p} - i_{D3n}, \tag{3.71}$$

where

$$i_{D3p} = i_{D1p}, \quad i_{D3n} = i_{D1n}. \tag{3.72}$$

From the above three equations, we obtain

$$i_{out+}(n-1) + i_{out-}(n) = -i_{in-}(n-1). \tag{3.73}$$

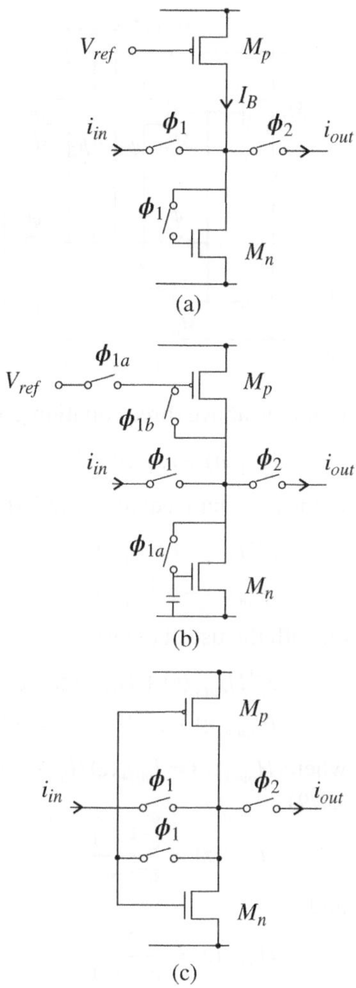

Figure 3.6 Second generation memory cells: basic memory cell (a), S^2I cell (b) and cell with $i_{in}/2$ memorized in each transistor (c).

On phase Φ_1, corresponding to the nth sampling period with the open switches driven by Φ_2, the transistors M_{n1}, M_{p1}, M_{n3}, and M_{p3} compose the current mirror with the output current

$$i_{out+}(n) = -[i_{in+}(n) + i_{Dp2} - i_{Dn2}]. \tag{3.74}$$

The output current i_{out-} can be calculated as:

$$i_{out-}(n-1) = i_{D4p} - i_{D4n}, \tag{3.75}$$

where

$$i_{D4p} = i_{D2p}, \quad i_{D4n} = i_{D2n}. \tag{3.76}$$

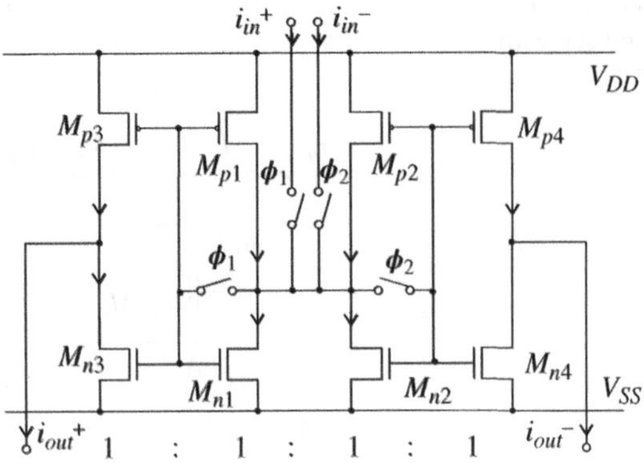

Figure 3.7 Circuit based on a second-generation memory cell.

From the above three equations, we obtain

$$i_{out+}(n) + i_{out-}(n-1) = -i_{in+}(n). \tag{3.77}$$

In the z domain, equations (3.73) and (3.77) have the form

$$\begin{aligned} &z^{-1}I_{out+}(z) + I_{out-}(z) = -z^{-1}I_{in-}(z) \\ &I_{out+}(z) + z^{-1}I_{out-}(z) = -I_{in+}(z), \end{aligned} \tag{3.78}$$

or, with the use of (3.69)

$$\begin{aligned} &z^{-1}H_{out+}(z) + H_{out-}(z) = z^{-1} \\ &H_{out+}(z) + z^{-1}H_{out-}(z) = -1, \end{aligned} \tag{3.79}$$

where $H_{out+}(z) = I_{out+}(z)/I_{in}(z)$ and $H_{out-}(z) = I_{out-}(z)/I_{in}(z)$. As the solution of (3.79),

$$H_{out+}(z) = \frac{z^{-2}+1}{z^{-2}-1}, \tag{3.80}$$

and

$$H_{out-}(z) = \frac{-2z^{-1}}{z^{-2}-1}, \tag{3.81}$$

are obtained.

The balanced integrator shown in Fig. 3.5 is composed of the two circuits presented in Fig. 3.7 with opposite signs of input currents delivered to each circuit. The currents from the inverting and noninverting outputs of the circuits are alternately added, giving cancellation of parasitic effects, and the transfer functions

$$H_{+}(z) = H_{out+} - H_{out-} = \frac{z^{-1}+1}{z^{-1}-1}, \tag{3.82}$$

and

$$H_{-}(z) = -H_{out+} + H_{out-} = -\frac{z^{-1}+1}{z^{-1}-1}, \tag{3.83}$$

for positive and negative paths, respectively.

An ideal bilinear integrator has the transfer function

$$H(z) = \pm k\frac{1+z^{-1}}{1-z^{-1}}. \tag{3.84}$$

The sign $\pm$ depends on the output, inverting or non-inverting, which is chosen. According to the bilinear transformation (2.3), the integrator described by (3.84) is the counterpart of an analog integrator with the transfer function $H^a(s) = 1/s$, where the factor $k = T/2$, or $k = 1/2$ for the sampling period $T = 1$ (Handkiewicz, 2002). Equations (3.82) and (3.83) denote that the integrator in Fig. 3.5 is the bilinear balanced integrator with $k = 1$. This value of the coefficient k denotes that the bilinear transformation described by (2.8), (2.16), and (2.17), with the sampling period $T = 2$, is realized by the integrator. As was mentioned, the ratios $1:1:1:1$ at the bottom of Fig. 3.7 denote that the dimensions of the channels in the input transistors, M_{n1} and M_{n2} are equal to the channel dimensions in the output transistors, M_{n3} and M_{n4}. The p-channel transistors satisfy the same condition. We can realize the bilinear integrator with the factor $k = 1/2$, taking the channels in the output transistors twice as narrow or long as in the input transistors.

3.8 Differential Stage

Let us consider the circuit presented in Fig. 3.8 and called a differential stage.

The transistor M_{nB} in this circuit plays the role of a current source that delivers the bias current I_B.

Assuming that the transistors M_{n1} and M_{n2} are in the saturated mode and have the same transconductances $\beta_{n1} = \beta_{n2}$, we can express the drain currents as

$$I_{D1} = \frac{\beta_n}{2}(V_{GS1} - V_{Tn})^2, \tag{3.85}$$

and

$$I_{D2} = \frac{\beta_n}{2}(V_{GS2} - V_{Tn})^2. \tag{3.86}$$

Hence, the difference voltage between the inputs which is defined by

$$V_d = V_1 - V_2 = V_{GS1} - V_{GS2}, \tag{3.87}$$

can be written in the form

$$V_d = \sqrt{\frac{2I_{D1}}{\beta_n}} - \sqrt{\frac{2I_{D2}}{\beta_n}}, \tag{3.88}$$

or

$$\frac{\beta_n}{2}V_d^2 = I_{D1} + I_{D2} - 2\sqrt{I_{D1}I_{D2}}. \tag{3.89}$$

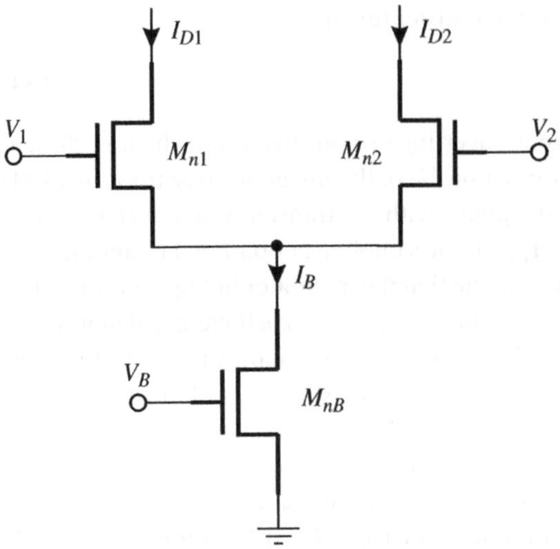

Figure 3.8 Differential stage.

From this equation, and from

$$I_{D1} + I_{D2} = I_B, \tag{3.90}$$

we can express the drain currents with the use of the difference voltage as

$$I_{D1} = \frac{I_B}{2} \pm \frac{1}{2}\sqrt{\beta_n I_B} V_d \sqrt{1 - \frac{\beta_n}{4I_B} V_d^2}, \tag{3.91}$$

and

$$I_{D2} = \frac{I_B}{2} \mp \frac{1}{2}\sqrt{\beta_n I_B} V_d \sqrt{1 - \frac{\beta_n}{4I_B} V_d^2}, \tag{3.92}$$

The plots of these currents are shown in Fig. 3.9.

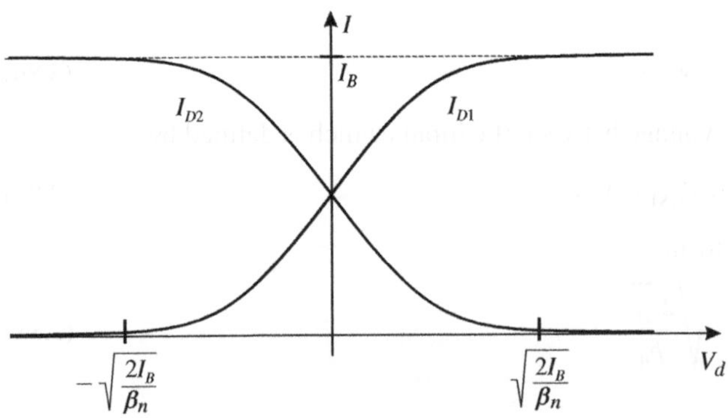

Figure 3.9 Currents in a differential stage.

3.9 Differential Transconductance Amplifier

A circuit composed of a differential stage and a current mirror with p-channel transistors, presented in Fig. 3.10, has the output current obtained as a difference of drain currents (3.91) and (3.92), in the form

$$I_{out} = I_{D1} - I_{D2} = \sqrt{\beta_n I_B} V_d \sqrt{1 - \frac{\beta_n}{4I_B} V_d^2}, \tag{3.93}$$

Assuming that in differential amplifiers

$$V_d \ll 2\sqrt{\frac{I_B}{\beta_n}}, \tag{3.94}$$

we can now write relation (3.93) in the following way

$$I_{out} = -\sqrt{\beta_n I_B}(V_2 - V_1), \tag{3.95}$$

where

$$g_m = \sqrt{\beta_n I_B}, \tag{3.96}$$

is the differential-mode transconductance of a differential OTA. The first input is called the noninverting one whereas the second is the inverting one.

It is possible to realize fully differential transconductance amplifiers with two outputs, inverting and noninverting (Fig. 3.11). Similarly, as it was in the case of

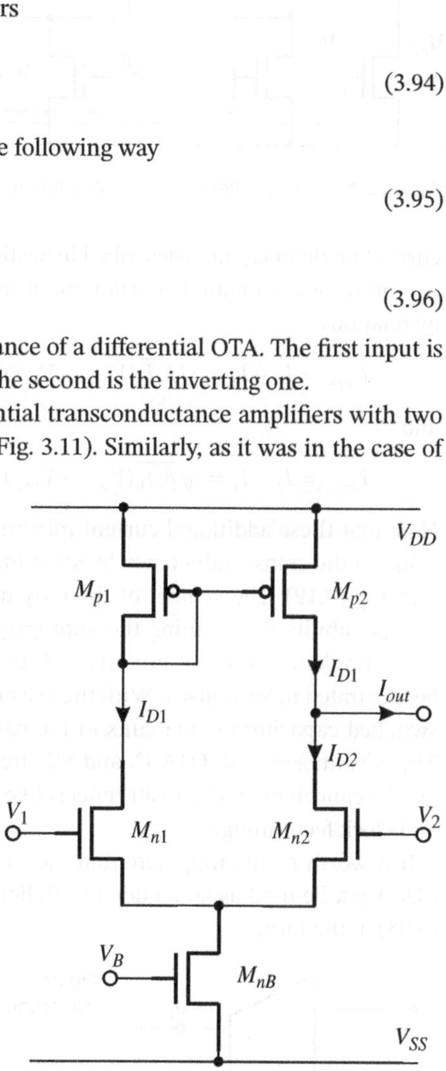

Figure 3.10 Differential transconductance amplifier.

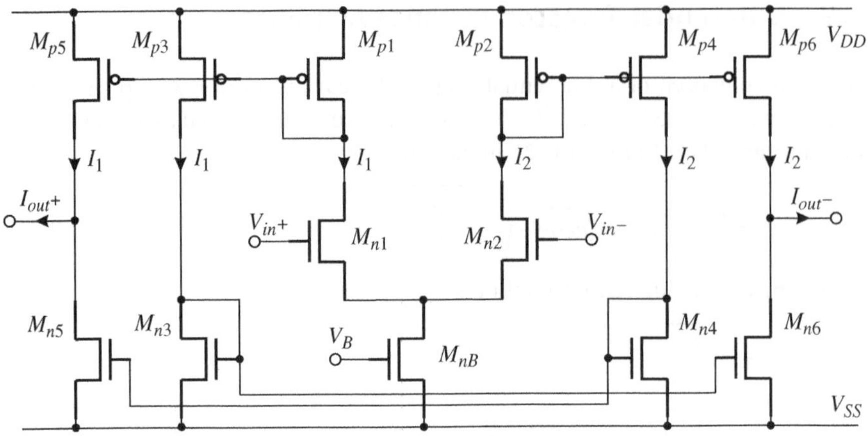

Figure 3.11 Fully differential transconductance amplifier.

current mode integrators described in Sections 3.6 and 3.7, additional current mirrors allow one to obtain two balanced outputs with the output currents are given by relations

$$I_{out+} = I_1 - I_2 = \sqrt{\beta_n I_B}(V_{in+} - V_{in-}), \tag{3.97}$$

and

$$I_{out-} = I_2 - I_1 = \sqrt{\beta_n I_B}(V_{in-} - V_{in+}), \tag{3.98}$$

Note that these additional current mirrors make it possible to give the required value to the transconductance by selecting the channel sizes, as shown in relationship (3.19). The symbol of the fully differential OTA is shown in Fig. 3.12. The possibility of obtaining the appropriate transconductance value allows the direct implementation of prototype gC circuits in the OTA-C technique, as will be illustrated in Section 4.1. With the use of fully differential amplifiers, so-*called switched capacitor (SC) circuits in the balanced structure can also be obtained. The advantages of SI, OTA-C, and SC circuits realized in the balanced structure are the cancellation of parasitic effects like dc offset, cross talk from digital signals and clock feedthrough.

It is worth mentioning here that the transconductance amplifier described by (3.95) can be used as an analog multiplier. On the basis of (3.5), we can rewrite (3.95) in the form

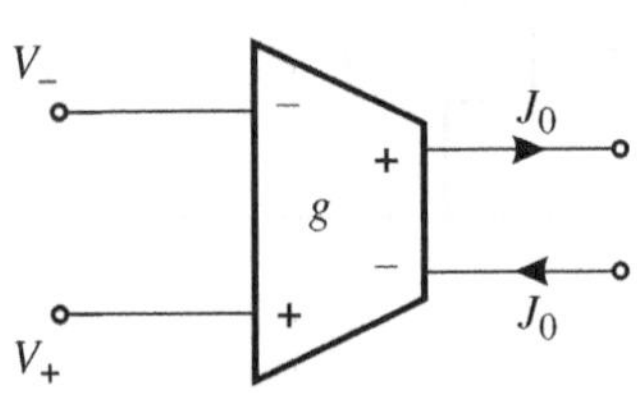

Figure 3.12 Symbol of the fully differential transconductance amplifier.

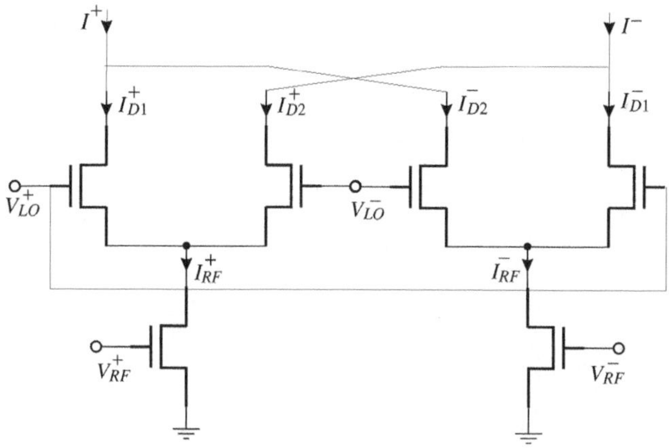

Figure 3.13 Mixer cell.

$$I_{out} = -\frac{\beta_n}{\sqrt{2}}(V_B - V_{Tn})(V_2 - V_1). \tag{3.99}$$

Two differential stages from Fig. 3.8 that are connected in a cross-coupled balanced structure compose the Gilbert cell (Gilbert, 1968). The Gilbert cell is widely used to multiply a radio frequency (RF) and a local oscillator (LO) signals in digital wireless personal communication systems as so-called mixer. The schematic diagram of a mixer is shown in Fig. 3.13. Similarly, as it was in the case of transconductance amplifiers, the drain currents in this cell can be written as follows:

$$\begin{aligned}
I_{D1}^{+} &= \frac{I_{RF}^{+}}{2} + \frac{\beta_n}{2\sqrt{2}}(V_{RF}^{+} - V_{Tn})(V_{LO}^{+} - V_{LO}^{-}),\\
I_{D2}^{+} &= \frac{I_{RF}^{+}}{2} - \frac{\beta_n}{2\sqrt{2}}(V_{RF}^{+} - V_{Tn})(V_{LO}^{+} - V_{LO}^{-}),\\
I_{D1}^{-} &= \frac{I_{RF}^{-}}{2} + \frac{\beta_n}{2\sqrt{2}}(V_{RF}^{-} - V_{Tn})(V_{LO}^{+} - V_{LO}^{-}),\\
I_{D2}^{-} &= \frac{I_{RF}^{-}}{2} - \frac{\beta_n}{2\sqrt{2}}(V_{RF}^{-} - V_{Tn})(V_{LO}^{+} - V_{LO}^{-}).
\end{aligned} \tag{3.100}$$

Denoting $I_{RF}^{+} = -I_{RF}^{-} = I_{RF}$ $V_{RF}^{+} = -V_{RF}^{-} = V_{RF}$ and $V_{LO}^{+} = -V_{LO}^{-} = V_{LO}$ we obtain:

$$\begin{aligned}
I^{+} &= I_{D1}^{+} + I_{D2}^{-} = \sqrt{2}\beta_n V_{RF} V_{LO},\\
I^{-} &= I_{D1}^{-} + I_{D2}^{+} = -\sqrt{2}\beta_n V_{RF} V_{LO}.
\end{aligned} \tag{3.101}$$

Equation (3.101) show that the mixer is realized as the RF transconductor in which output currents I^{+}, I^{-} are commutated by the LO.

4

Hardware Implementation of Mixed Signal Systems

The chapter presents new perspectives in the design of filters based on multiport prototype gyrator–capacitor (gC) circuits. First, the design of a filter with corrected phase characteristics is presented. Since the transfer function of such a filter has zeros located in the right half-plane of the complex variable s, it cannot be implemented using a prototype reciprocal circuit, such as the ladder reactance network. The nonreciprocal gC circuit is used for this purpose. This prototype circuit was implemented in the OTA-C technique, using the fully differential transconductance amplifiers described in Section 3.9.

The use of one-dimensional (1D), multiport gC circuits to design filter banks is presented in Sections 4.1, 4.2, 4.4 and 4.6. The use of filter banks is demonstrated in an example of a frequency-interleaved (FI) analog-to-digital converter (ADC). The analysis filter bank of this converter is implemented as a switched-current (SI) circuit. However, the gC prototype circuit of the synthesis filter bank in the converter is implemented as a digital circuit.

4.1 Filter Design in OTA-C Technique

To illustrate the synthesis method based on the prototype gC circuit, let us consider a low-pass filter with the transfer function

$$H^a(s) = \frac{s^2 - 2s + 3}{(s^2 + 2s + 3)(s + 1)} = \frac{P'S}{Q'S}, \tag{4.1}$$

where

$$P' = [1\ \ -2\ \ 3],\quad Q' = [1\ \ 3\ \ 5\ \ 3]. \tag{4.2}$$

The above transfer function contains an all-pass block introduced to improve the linearity of the phase characteristic. The magnitude and phase characteristics are shown in Fig. 4.1. The amplitude response is shown in Fig. 4.1a, assuming

Methods and Tools for Computer-aided Design of Multidimensional Filters in Signal Processing,
First Edition. Andrzej Handkiewicz.

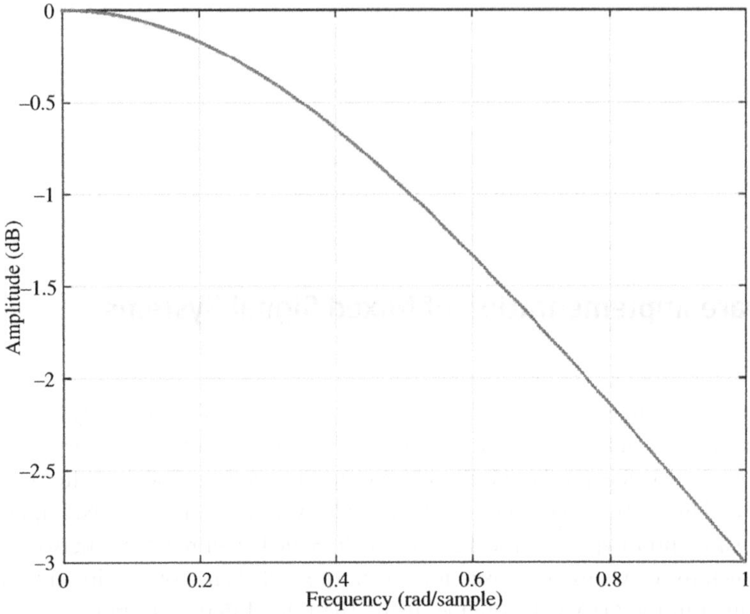

(a)

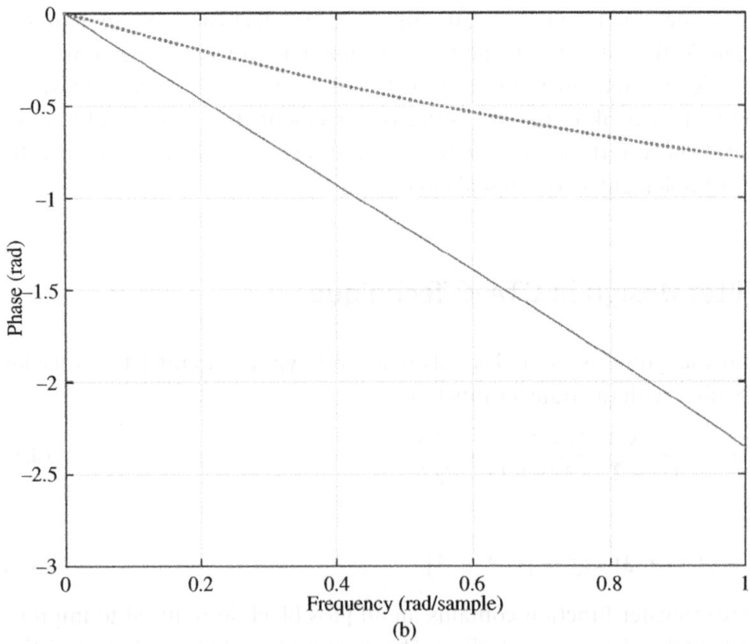

(b)

Figure 4.1 Amplitude (a) and phase (b) responses of the filter described by the transfer function (4.1).

that the cut-off frequency corresponds to 3 dB attenuation. In Fig. 4.1b, the phase characteristics in this band are shown without and with the linearity improvement, denoted by the dotted and solid lines, respectively.

The numerator of the transfer function of a ladder reactance filter must be a purely even or odd polynomial (Temes and LaPatra, 1977). Hence, transfer function (4.1) cannot be obtained on the basis of the ladder reactance structure. However, this transfer function fulfils the relation $|H(j\omega)| \leq 1$ and, in accordance with (1.25), (1.27), and (1.28), can be realized by a lossless non-reciprocal two-port network terminated in the conductances G_i and G_o. Such gC circuit described in VHDLA as

```
library VLSI;
use VLSI.SI.all;
entity example is port (
        terminal input Vin : electrical;
        terminal output Vout : electrical;
        terminal ground gnd : electrical);
end entity example;
-- some comment line
architecture example_arch of example is
variable v1,v2,v3 : real;
begin
-- capacitances
C : GC_CAP generic map(dim=>1) port map (Ni=>v2, Nj=>gnd);
Ci : GC_CAP generic map(dim=>1) port map (Ni=>v1, Nj=>gnd);
Co : GC_CAP generic map(dim=>1) port map (Ni=>v3, Nj=>gnd);
Cf : GC_CAP generic map(dim=>1) port map (Ni=>v1, Nj=>v3);
-- conductances with I/O signals
Gi : GC_CON port map (Ni=>Vin, Nj=>v1);
Go : GC_CON port map (Ni=>v3, Nj=>gnd, Vo=>Vout);
-- gyrators
ig1 : GC_GYR port map( Ni=>v1, Nj=>v2);
ig2 : GC_GYR port map( Ni=>v2, Nj=>v3);
igf : GC_GYR port map( Ni=>v1, Nj=>v3);

end architecture;
```

is presented in Fig. 4.2a. In the **gCstudio** environment, the file **PhaseL4C.sn** containing results of the symbolic analysis of this circuit is generated. Running **optimizer**

```
for ((i=0; i<=29; i++));
do
./optimizer -c 1 -x -n -t 0.0001 -a s -i 10000 PhaseL4C.sn init4C >
                                                                  "f$i"
grep Goal "f$i"
done
```

allows to obtain, at the goal function value 0.00000000000114, the following values of the parameters of the gC circuit:

```
1.013920569284684
0.058023187349413
1.294223738975079
0.673823382280111
1.492012995596362
1.373721397579230
-1.347646779973474
1.463693152381484
0.106845642140067
```

The **initC4** file is as follows:

```
0.5 0 1.5 0.000001
0.5 0 1.5 0.000001
0.5 0 1.5 0.000001
0.5 0 1.5 0.000001
0 -1.5 1.5 0.000001
0 -1.5 1.5 0.000001
0 -1.5 1.5 0.000001
0.5 0 1.5 0.000001
0.5 0 1.5 0.000001
```

and the files **s0** and **s1**, with coefficients given in (4.1), are:

```
1
3
0 3
1 5
2 3
3 1
```

and

```
1
3
0 3
1 -2
2 1
3 0
```

respectively.

The circuit presented in Fig. 4.2a can be directly implemented as an OTA-C filter with the use of two transconductance amplifiers for each gyrator and two additional amplifiers for input and output conductances G_1 and G_2.

The fully differential transconductance amplifier presented in Fig. 3.11, with the symbol shown in Fig. 3.12, can be used in such an implementation. The output currents I_o are proportional to the difference of voltages V_+ and V_-, $I_o = g(V_+ - V_-)$, where g is the amplifier transconductance. Hence, the ideal gyrator in Fig. 4.3a, whose input current $I_i = gV_o$ and output current $I_o = gV_i$, as is described by admittance matrix (1.1), is realized with the use of two transconductance

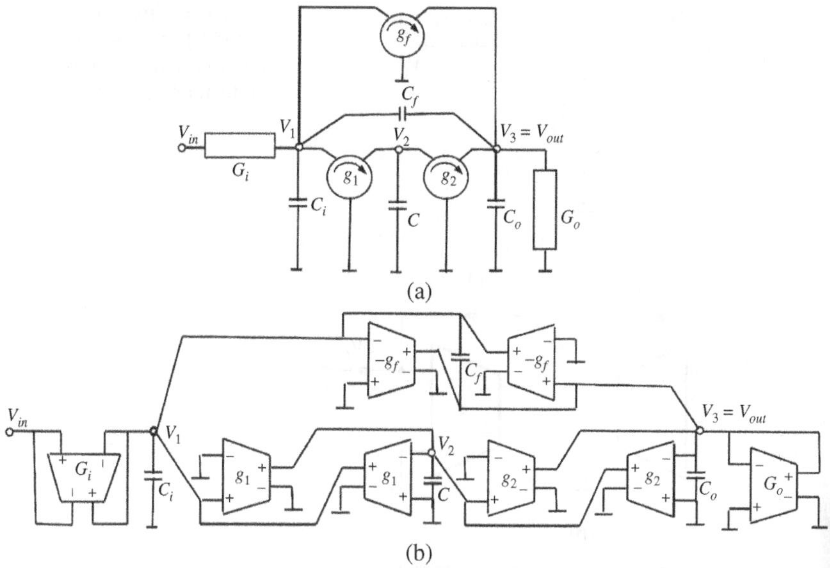

Figure 4.2 OTA-C implementation (b) of the gyrator–capacitor circuit (a) described by transfer function (4.1).

amplifiers, as shown in Fig. 4.3b. The ideal gyrator realized with the use of transconductance amplifiers in the balanced structure is presented in Fig. 4.3c. Implementations of floating and grounded resistors with conductances G are given in Fig. 4.4.

On the basis of gyrators and resistors implemented with the use of transconductance amplifiers, it is easy to obtain the OTA-C filter shown in Fig. 4.2b, which is the counterpart network of the gC filter shown in Fig. 4.2a.

4.2 Design of Filter Banks

Filter banks are necessary to use in a wide range of applications, from data transmission to data compression and signal processing, for example to speech, audio, video, and biosignal coding (Szczesny et al., 2012; Raychaudhuri and Mandayam, 2012; Magalhaes et al., 2013; Simón Gálvez et al., 2015). Sections 4.4–4.6, present **gCstudio**, **NANOstudio**, and **SIstudio** environments for designing filter banks in terms of their implementation in standard, digital nanotechnologies. The functioning of this environments has been illustrated by its application for FI ADC. The presented design strategy uses multiport lossless gC circuits as prototype networks, both for analysis and synthesis filters in FI ADC.

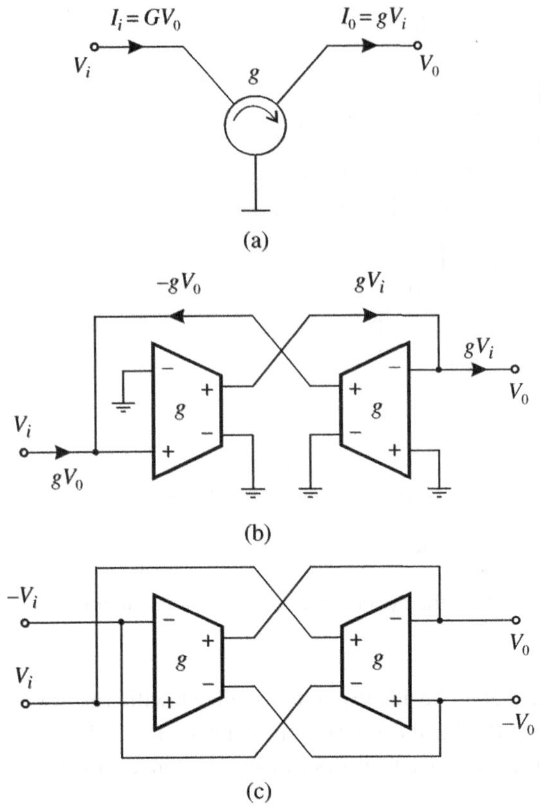

Figure 4.3 Implementations (b) and (c) of ideal gyrator (a) with the use of transconductance amplifiers.

4.3 General Structure of FI ADC

General structure of FI ADC is presented in Fig. 4.5. In this section, we will focus on filter banks H_i and G_i, $i = 1, \ldots, M$, while blocks Q, which perform quantization and decimation of signals, are omitted in our considerations. The essential difference versus the ones so far presented in the literature Zhao and Chan (2009), Qiu et al. (2014), and Läwenborg et al. (2003), lies in replacement of analysis and synthesis filters, described by transfer functions in the analog domain, respectively, as

$$H_1^a(s),\ H_2^a(s),\ \ldots,\ H_M^a(s), \tag{4.3}$$

and

$$G_1^a(s),\ G_2^a(s),\ \ldots,\ G_M^a(s), \tag{4.4}$$

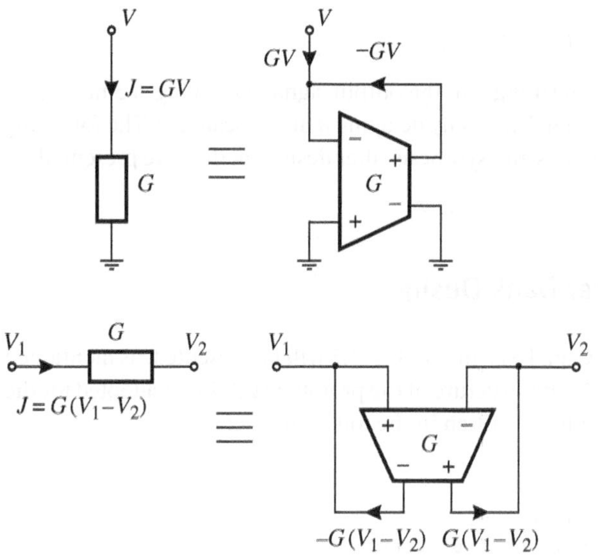

Figure 4.4 Implementation of conductances based on transconductance amplifiers.

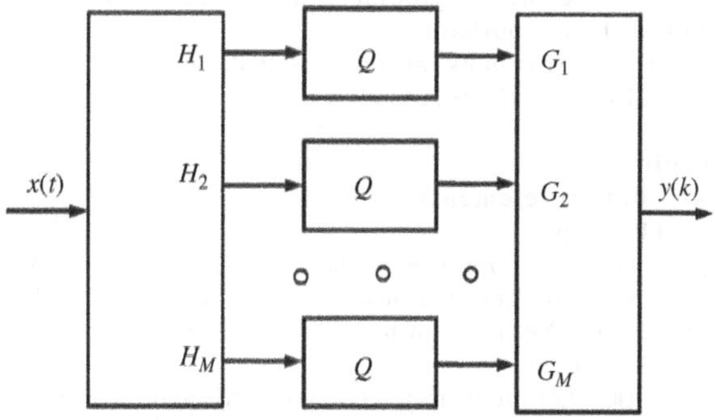

Figure 4.5 Structure of an M-channel FI ADC composed of multiport analysis and synthesis filters.

present separately in each converter path, with multiport networks. Similarly as in Section 2.3, we will also use these transfer functions in the discrete z domain using bilinear transformation (2.3), then leaving out upper index a in their marking.

The applied FI ADC design methods use transfer functions (4.3) and (4.4) of the complex variable s only for filter banks composed of two-port networks or filter pairs, following the approach presented in Sections 1.2 and 1.3. In this section, we will show you how to design multiport analysis and synthesis filter banks using their time domain description. This possibility is provided by the perfect reconstruction (PR) condition for FI ADC, which in the time domain is described by the relation

$$y(k) = cx(k - d), \quad k = 0, 1, 2, \ldots \tag{4.5}$$

(Läwenborg et al., 2003) meaning that the output signal $y(k)$ in Fig. 4.5 may differ from the sampled input signal $x(k)$ only delaying d and rescaling c. The following examples of this new analysis and synthesis filter design method are presented for the case $M = 3$.

4.4 Analysis Filter Bank Design

The design of the FI ADC analysis filter in Fig. 4.5 will be presented, as mentioned above, for the case $M = 3$. The structure of the prototype gC circuit adopted for the implementation of this filter is written in the following file:

```
library VLSI;
use VLSI.SI.all;
entity 3outBank is port (
        terminal output VoL : electrical;
        terminal output VoC : electrical;
        terminal output VoH : electrical;
        terminal input Vin : electrical;
        terminal ground gnd : electrical);
end entity 3outBank;
architecture bank_arch of 3outBank is
variable v1,v2,v3,v4,v5,v6,v7 : real;

begin
--- output elements
-- LP output
G1 : GC_CON port map (Ni=>v7, Nj=>gnd, Vo=>VoL);
C1 : GC_CAP generic map(dim=>1) port map (Ni=>v7, Nj=>gnd);
ig1 : GC_GYR port map( Ni=>v4, Nj=>v7);
-- CP output
G2 : GC_CON port map (Ni=>v6, Nj=>gnd, Vo=>VoC);
C2 : GC_CAP generic map(dim=>1) port map (Ni=>v6, Nj=>gnd);
ig2 : GC_GYR port map( Ni=>v4, Nj=>v6);
-- HP output
G3 : GC_CON port map (Ni=>v5, Nj=>gnd, Vo=>VoH);
ig3 : GC_GYR port map( Ni=>v4, Nj=>v5);
--- chain elements
ig4 : GC_GYR port map( Ni=>v1, Nj=>v2);
C4 : GC_CAP generic map(dim=>1) port map (Ni=>v2, Nj=>gnd);
ig5 : GC_GYR port map( Ni=>v2, Nj=>v3);
C5 : GC_CAP generic map(dim=>1) port map (Ni=>v3, Nj=>gnd);
ig6 : GC_GYR port map( Ni=>v3, Nj=>v4);
C6 : GC_CAP generic map(dim=>1) port map (Ni=>v4, Nj=>gnd);
-- input node
G4 : GC_CON port map (Ni=>Vin, Nj=>v1);
```

```
-- LP output
ig11 : GC_GYR port map( Ni=>v1, Nj=>v7);
ig12 : GC_GYR port map( Ni=>v2, Nj=>v7);
ig13 : GC_GYR port map( Ni=>v3, Nj=>v7);
-- CP output
ig21 : GC_GYR port map( Ni=>v1, Nj=>v6);
ig22 : GC_GYR port map( Ni=>v2, Nj=>v6);
ig23 : GC_GYR port map( Ni=>v3, Nj=>v6);
-- HP output
ig31 : GC_GYR port map( Ni=>v1, Nj=>v5);
ig32 : GC_GYR port map( Ni=>v2, Nj=>v5);
ig33 : GC_GYR port map( Ni=>v3, Nj=>v5);

end architecture;
```

named **1in3o1D.vhdla**.

The description above shows that the prototype circuit has one input (*input node*) with conductance **G4** assigned to it, and three outputs: the low-pass (*LP*), the band-pass (*CP*), and the high-pass (*HP*), terminated in the conductances **G1**, **G2**, and **G3**, respectively. The capacitances **C1** and **C2** are additionally connected to the low-pass and band-pass outputs. All outputs are connected to node **v4** with gyrators **ig1**, **ig2**, **ig3**. The nodes **v1**, **v2**, **v3**, **v4** are connected in pairs by gyrators **ig4**, **ig5**, **ig6** forming a chain connecting input node **v1** with the node **v4**. The nodes **v2**, **v3**, **v4** are connected to the properly grounded capacitances **C4**, **C5**, **C6**, and the nodes **v1**, **v2**, **v3**, through the appropriate gyrators, to all three outputs.

The values of the gC prototype circuit parameters listed above can be found in the **NANOstudio** environment:

```
for ((i=0; i<=9; i++)); do
./optimizer -c 1 -x -n -t 0.01 -a m -i 1000 -d 16 1 1 1in3o1D.sn
                                                   init1D > "f$i"
   grep Goal "f$i"
   done
```

The file **1in3o1D.sn** is generated in the **gCstudio** environment based on the description of the gC prototype circuit given above, while the initialization file **init1D** is as follows:

```
0.5 0 1.5 0.000001
0.5 0 1.5 0.000001
0.5 0 1.5 0.000001
0.5 0 1.5 0.000001
0.5 0 1.5 0.000001
0 -1.5 1.5 0.000001
0 -1.5 1.5 0.000001
0 -1.5 1.5 0.000001
0 -1.5 1.5 0.000001
0 -1.5 1.5 0.000001
0 -1.5 1.5 0.000001
```

```
0 −1.5 1.5 0.000001
0 −1.5 1.5 0.000001
0 −1.5 1.5 0.000001
0 −1.5 1.5 0.000001
0 −1.5 1.5 0.000001
0 −1.5 1.5 0.000001
0 −1.5 1.5 0.000001
0 −1.5 1.5 0.000001
0 −1.5 1.5 0.000001
0.5 0 1.5 0.000001
0.5 0 1.5 0.000001
0.5 0 1.5 0.000001
0.5 0 1.5 0.000001
```

and such a notation means that values of conductance and capacitance parameters are searched in the range from 0 to 1.5, and transconductances of gyrators in the range from −1.5 to 1.5.

As it is known from **notes** placed in the description of the **NANOstudio** environment in Section 1.1.2, the analysis option *m* used in **./optimizer** means that the designed analysis filter bank has connected discrete time circuits to all three outputs. At this stage of FI ADC design, they are given by three transfer functions with numerator and denominator polynomials equal simply to 1, described by the corresponding files **zd1**, **zd2**, **zd3**, **zn1**, **zn2**, **zn3**, all with the content:

```
1
0
0          1.0
```

The files **t1**, **t2**, **t3**, containing the expected signal values at the low-pass, band-pass, and high-pass outputs of the prototype circuit after its bilinear transformation (2.3), are in turn given as the output signals of the appropriate fourth order elliptic filters. For this goal, the following file can be used in Octave:

```
%Responses of the 4-th order filter bank
pkg load video
pkg load signal
clear all;

%impulse responses
t0=[1 0 0 0 0 0 0 0 0 0 0 0 0 0 0 0];
[nL,dL]=ellip(4,1,20,0.3);
yL=filter(nL,dL,t0);
[nC,dC]=ellip(4,1,20,[0.3, 0.6]);
yC=filter(nC,dC,t0);
[nH,dH]=ellip(4,1,20,0.6,'high');
yH=filter(nH,dH,t0);

fid=fopen('t1','w');
fprintf(fid,'%f\n',yL);
```

```
fclose(fid);
fid=fopen('t2','w');
fprintf(fid,'%f\n',yC);
fclose(fid);
fid=fopen('t3','w');
fprintf(fid,'%f\n',yH);
fclose(fid);
```

for the Kronecker delta appearing in the input signal **t0** as the first sample.

As one of the solutions, for the value of the objective function equal to 0.080465214292211, the following values of the parameters of the three-output filter bank were obtained:

```
1.409594156225392
1.111631506706062
0.696502403132663
0.333040223942030
0.307795999349264
0.234920948465562
-0.933298167152449
0.000823552669232
1.245762715017640
-1.487989888840648
-0.170094898439695
1.370809755422309
-0.479063698119707
0.803854773153754
0.619892041685004
-0.186695586985271
-0.401749058538908
0.009474288793113
-0.004545284632322
0.008067511387891
0.035910853158513
0.000001157580398
0.010130514478924
0.934426603654994
```

with the achieved frequency characteristics in the discrete domain, shown in Fig. 4.6. These characteristics were calculated in the Octave environment for the transfer function of the designed analysis filter bank with polynomial coefficients of the denominator (**z0**):

```
1
5
0          1.21161437
1         -0.33489478
2          0.89944786
3         -0.35680836
4          0.25588834
5         -0.05869171
```

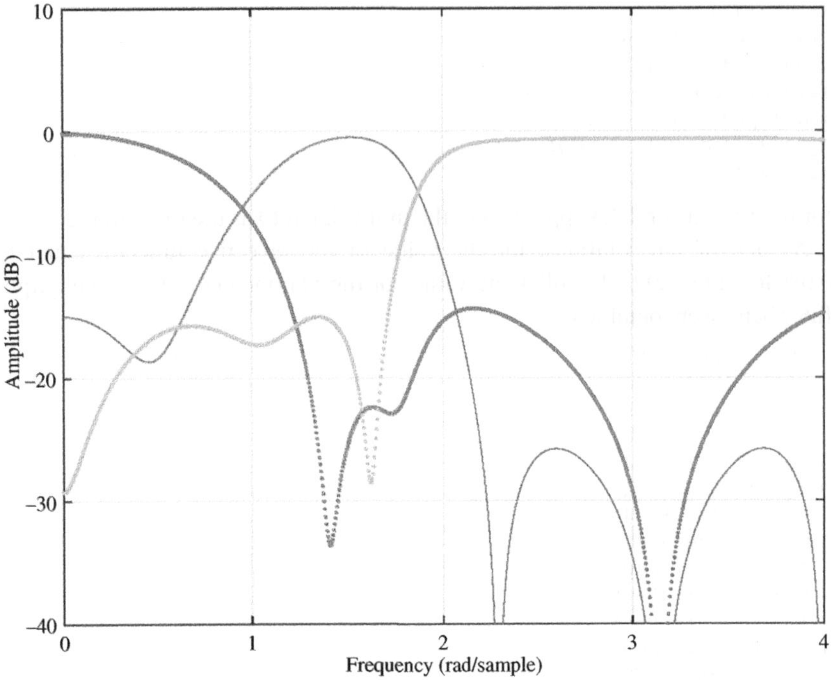

Figure 4.6 Frequency characteristics of the three-output analysis filter bank.

and numerators, for the low-pass output (**z1**):

1	
5	
0	0.17285106
1	0.18715948
2	0.38892978
3	0.37907827
4	0.23332599
5	0.22886905

for the band-pass output (**z2**):

1	
5	
0	0.13256061
1	0.02925781
2	−0.14965941
3	−0.07774471
4	0.16097078
5	0.19235888

and for the high-pass output (**z3**):

```
1
5
0     0.18572795
1    -0.41004220
2     0.69469130
3    -0.72065526
4     0.53739363
5    -0.34285733
```

respectively.

The above files with transfer function coefficients as well as the impulse response **tOut** shown in Fig. 4.9 (samples marked with crosses) were obtained after running the **Fil** filter:

```
./Fil 1in3o1D.sn fBankD13e4 16 1 1
```

where **1in3o1D.sn** is the previously used result of the symbolic analysis of the circuit with the structure **1in3o1D.vhdla** and **fBankD13e4** contains the given above values of the filter parameters. The last three arguments of the program **Fil** mean that the synthesized circuit is 1D and contains 16 samples in the first dimension. The **Fil** program is the result of the filter **filter1.cpp** compilation. The filter is designed in C^{++} using the members of the classes described in Sections 2.2 and 2.4. Both the program **filter1.cpp** and its compiled version **Fil**, along with the classes used, are available in the directory *Filter*.

4.5 Analysis Filter Bank Implementation in SI Technique

The analysis filter is, in our design method, a discrete time analog circuit. It can therefore be implemented by one of two techniques: switched capacitors (SC) or SI. We will use in this example the latter technique, while also illustrating the **SIstudio** environment mentioned in Section 3.1. The use of this environment requires extending the description of the prototype gc circuit of the analysis filter with the values of its elements:

```
library VLSI;
use VLSI.SI.all;
entity 3outBank is port (
        terminal output VoL : electrical;
        terminal output VoC : electrical;
        terminal output VoH : electrical;
        terminal input Vin : electrical;
        terminal ground gnd : electrical);
end entity 3outBank;
```

```
architecture bank_arch of 3outBank is
variable v1,v2,v3,v4,v5,v6,v7 : real;
-- input vertices
variable v8 : real;

constant c1 :    real :=  1.499949894081704;
constant c2 :    real :=  1.219593110468921;
constant c4 :    real :=  0.461643327162513;
constant c5 :    real :=  0.379374921380441;
constant c6 :    real :=  0.724814595288430;
constant IG1 :   real := -0.383272404312882;
constant IG2 :   real :=  1.499969856260605;
constant IG3 :   real := -0.001643471236039;
constant IG4 :   real := -1.072689433490010;
constant IG5 :   real :=  1.287891335381647;
constant IG6 :   real :=  0.284477888037863;
constant IG11 :  real :=  1.473827329879955;
constant IG12 :  real :=  0.412859023822297;
constant IG13 :  real :=  0.877705969812803;
constant IG21 :  real :=  0.653695375249911;
constant IG22 :  real :=  0.154140608228301;
constant IG23 :  real := -0.470735700448479;
constant IG31 :  real :=  0.011839319235986;
constant IG32 :  real :=  0.004410599807538;
constant IG33 :  real :=  0.010225199159283;
constant g1 :    real :=  0.059946542558399;
constant g2 :    real :=  0.000001586739711;
constant g3 :    real :=  0.012628152408397;
constant g4 :    real :=  1.027029194107652;

-- components
component IG is
generic( transcon : real);
port( terminal input x: electrical;
terminal output y : electrical;
terminal ground gnd : electrical);
end component IG;


component C is
generic( capacity : real);
port( terminal x : electrical;
terminal y : electrical);
end component C;


component G is
generic( conduct : real);
port( terminal x : electrical;
terminal y : electrical);
end component G;
```

```
--------------------------------------------------------
begin
-- input signals
Vin <= v8;

-- output signals
VoL <= v7;
VoC <= v6;
VoH <= v5;

--- output elements
-- LP output
--G1 : GC_CON port map (Ni=>v7, Nj=>gnd, Vo=>VoL);
G1 :    G         generic map( conduct=>g1)
                  port map( x=>v7, y=>gnd);
--C1 : GC_CAP generic map(dim=>1) port map (Ni=>v7, Nj=>gnd);
C1 :    C         generic map( capacity=>c1)
                  port map( x=>v7, y=>gnd);
--ig1 : GC_GYR port map( Ni=>v4, Nj=>v7);
ig1 :   IG        generic map( transcon=>IG1)
                  port map( x=>v4, y=>v7, gnd=>gnd);
-- CP output
--G2 : GC_CON port map (Ni=>v6, Nj=>gnd, Vo=>VoC);
G2 :    G         generic map( conduct=>g2)
                  port map( x=>v6, y=>gnd);
--C2 : GC_CAP generic map(dim=>1) port map (Ni=>v6, Nj=>gnd);
C2 :    C         generic map( capacity=>c2)
                  port map( x=>v6, y=>gnd);
--ig2 : GC_GYR port map( Ni=>v4, Nj=>v6);
ig2 :   IG        generic map( transcon=>IG2)
                  port map( x=>v4, y=>v6, gnd=>gnd);
-- HP output
--G3 : GC_CON port map (Ni=>v5, Nj=>gnd, Vo=>VoH);
G3 :    G         generic map( conduct=>g3)
                  port map( x=>v5, y=>gnd);
--ig3 : GC_GYR port map( Ni=>v4, Nj=>v5);
ig3 :   IG        generic map( transcon=>IG3)
                  port map( x=>v4, y=>v5, gnd=>gnd);
--- chain elements
--ig4 : GC_GYR port map( Ni=>v1, Nj=>v2);
ig4 :   IG        generic map( transcon=>IG4)
                  port map( x=>v1, y=>v2, gnd=>gnd);
--C4 : GC_CAP generic map(dim=>1) port map (Ni=>v2, Nj=>gnd);
C4 :    C         generic map( capacity=>c4)
                  port map( x=>v2, y=>gnd);
--ig5 : GC_GYR port map( Ni=>v2, Nj=>v3);
ig5 :   IG        generic map( transcon=>IG5)
                  port map( x=>v2, y=>v3, gnd=>gnd);
--C5 : GC_CAP generic map(dim=>1) port map (Ni=>v3, Nj=>gnd);
C5 :    C         generic map( capacity=>c5)
                  port map( x=>v3, y=>gnd);
```

```
--ig6 : GC_GYR port map( Ni=>v3, Nj=>v4);
ig6 :   IG       generic map( transcon=>IG6)
                 port map( x=>v3, y=>v4, gnd=>gnd);
--C6 : GC_CAP generic map(dim=>1) port map (Ni=>v4, Nj=>gnd);
C6 :    C        generic map( capacity=>c6)
                 port map( x=>v4, y=>gnd);
-- input signal
--G4 : GC_CON port map (Ni=>Vin, Nj=>v1);
G4 :    G        generic map( conduct=>g4)
                 port map( x=>v8, y=>v1);
-- LP output
--ig11 : GC_GYR port map( Ni=>v1, Nj=>v7);
ig11 :  IG       generic map( transcon=>IG11)
                 port map( x=>v1, y=>v7, gnd=>gnd);
--ig12 : GC_GYR port map( Ni=>v2, Nj=>v7);
ig12 :  IG       generic map( transcon=>IG12)
                 port map( x=>v2, y=>v7, gnd=>gnd);
--ig13 : GC_GYR port map( Ni=>v3, Nj=>v7);
ig13 :  IG       generic map( transcon=>IG13)
                 port map( x=>v3, y=>v7, gnd=>gnd);
-- CP output
--ig21 : GC_GYR port map( Ni=>v1, Nj=>v6);
ig21 :  IG       generic map( transcon=>IG21)
                 port map( x=>v1, y=>v6, gnd=>gnd);
--ig22 : GC_GYR port map( Ni=>v2, Nj=>v6);
ig22 :  IG       generic map( transcon=>IG22)
                 port map( x=>v2, y=>v6, gnd=>gnd);
--ig23 : GC_GYR port map( Ni=>v3, Nj=>v6);
ig23 :  IG       generic map( transcon=>IG23)
                 port map( x=>v3, y=>v6, gnd=>gnd);
-- HP output
--ig31 : GC_GYR port map( Ni=>v1, Nj=>v5);
ig31 :  IG       generic map( transcon=>IG31)
                 port map( x=>v1, y=>v5, gnd=>gnd);
--ig32 : GC_GYR port map( Ni=>v2, Nj=>v5);
ig32 :  IG       generic map( transcon=>IG32)
                 port map( x=>v2, y=>v5, gnd=>gnd);
--ig33 : GC_GYR port map( Ni=>v3, Nj=>v5);
ig33 :  IG       generic map( transcon=>IG33)
                 port map( x=>v3, y=>v5, gnd=>gnd);

end architecture;
```

The parameter values are placed as constants in the file header. However, the description of individual elements, compared to the one given in Section 4.4, additionally includes these constants as their values. The previous way of describing circuit elements is left in the current file as a comment.

The implementation of the above gC circuit in the SI technique is fully automated by the **SIstudio** environment. The program **SImaker** available in this environment is run in two steps, as shown in the following batch program:

```
@echo off
title SI Maker
@echo on
C:\...\SIstudio\SImaker EQUE 1in3o1D.vhdla SI1in3o.equ
C:\...\SIstudio\SImaker SCHE SI1in3o.equ SI1in3o.vhdl
@echo off
pause
```

for Windows operating system.

In the first step, based on the above description of the gC circuit contained in the file **1in3o1D.vhdla**, the system of integral equations described in the file **SI1in3o.equ** is generated for the option *EQUE*. We assume that the batch program is located in a directory shared with these files, and the program **SImaker** is pointed to by the appropriate path. The integral equations, described in the file ***.equ**, are obtained by transforming the nodal equations of the gC circuit.

The second step is to run **SImaker** with the option *SCHE*. As a result, the SI network is generated, which is the counterpart network of the gC circuit. The SI circuit is described in the file **SI1in3o.vhdl** with the content:

```
library VLSI;
use VLSI.SI.all;
---------------------------------------
entity 3outBank is
        port(   terminal input Vin+, Vin- : electrical;
          terminal output VoC+, VoC- : electrical;
          terminal output VoH+, VoH- : electrical;
          terminal output VoL+, VoL- : electrical;
          terminal clock clk : electrical;
          terminal ground gnd : electrical);
end entity 3outBank;
---------------------------------------
---------------------------------------
architecture SCHEMATIC of 3outBank is

 constant const01 : real := 1.0;
 constant const02 : real := 1.044;
 constant const03 : real := 0.01153;
 constant const04 : real := 0.6365;
 constant const05 : real := 1.435;
 constant const06 : real := 2.324;
 constant const07 : real := 2.79;
 constant const08 : real := 0.009554;
 constant const09 : real := 0.3339;
 constant const10 : real := 0.8943;
 constant const11 : real := 3.395;
 constant const12 : real := 0.7499;
 constant const13 : real := 0.02695;
 constant const14 : real := 1.241;
 constant const15 : real := 2.314;
```

```
constant const16 : real := 0.3925;
constant const17 : real := 0.002267;
constant const18 : real := 2.069;
constant const19 : real := 0.5288;
constant const20 : real := 0.9375;
constant const21 : real := 0.3493;
constant const22 : real := 0.8097;
constant const23 : real := 0.1301;
constant const24 : real := 0.536;
constant const25 : real := 0.1264;
constant const26 : real := 0.386;
constant const27 : real := 1.23;
constant const28 : real := 1.301e-006;
constant const29 : real := 0.9826;
constant const30 : real := 0.2752;
constant const31 : real := 0.5852;
constant const32 : real := 0.2555;
constant const33 : real := 0.03997;

-- Integrator interface:
component INT is
   port( terminal input x+,x- : electrical;
         terminal output y+,y- : electrical;
         terminal clock clkint : electrical);
end component INT;

-- Current Mirror interface:
component CM is
   generic( coeff1 : real);
   port( terminal input x : electrical;
         terminal output y1 : electrical);
end component CM
component CM is
   generic( coeff1,coeff2,coeff3,coeff4 : real);
   port( terminal input x : electrical;
         terminal output y1,y2,y3,y4 : electrical);
end component CM
component CM is
   generic( coeff1,coeff2,coeff3,coeff4,coeff5 : real);
   port( terminal input x : electrical;
         terminal output y1,y2,y3,y4,y5 : electrical);
end component CM
component CM is
   generic( coeff1,coeff2,coeff3,coeff4,coeff5,coeff6 : real);
   port( terminal input x : electrical;
         terminal output y1,y2,y3,y4,y5,y6 : electrical);
end component CM

-- terminals:
terminal s01+,s01-, s1+,s1-, s2+,s2-, s3+,s3- : electrical;
terminal s4+,s4-, s5+,s5-, s6+,s6-, s7+,s7- : electrical;
```

```
terminal s8+,s8-, s9+,s9-, s10+,s10-, s11+,s11- : electrical;
terminal s12+,s12-, s13+,s13-, s14+,s14-, s15+,s15- : electrical;
terminal s16+,s16-, s17+,s17-, : electrical;

--------------------------------------------------------
begin

 -- input signals
s01+ <= Vin+; s01- <= Vin-;
 -- output signals
s15+ <= VoC+; s15- <= VoC-;
s16+ <= VoH+; s16- <= VoH-;
s17+ <= VoL+; s17- <= VoL-;

-- Integrators:
int1 : INT port map( x+=>s3+, x-=>s3-, y+=>s4+, y-=>s4-,
                                                  clkint=>clk);
int2 : INT port map( x+=>s5+, x-=>s5-, y+=>s6+, y-=>s6-,
                                                  clkint=>clk);
int3 : INT port map( x+=>s7+, x-=>s7-, y+=>s8+, y-=>s8-,
                                                  clkint=>clk);
int4 : INT port map( x+=>s9+, x-=>s9-, y+=>s10+, y-=>s10-,
                                                  clkint=>clk);
int6 : INT port map( x+=>s13+, x-=>s13-, y+=>s14+, y-=>s14-,
                                                  clkint=>clk);

-- Current Mirrors:
CM00p : CM generic map ( coeff1=>const01)
       port map( x=>s01+, y1=>s2-);
CM00m : CM generic map ( coeff1=>const01)
       port map( x=>s01-, y1=>s2+);

CM01p : CM generic map ( coeff1=>const06, coeff2=>const24,
                          coeff3=>const20, coeff4=>const29)
       port map( x=>s2+, y1=>s3+, y2=>s9-, y3=>s12-, y4=>s13-);
CM01m : CM generic map ( coeff1=>const06, coeff2=>const24,
                          coeff3=>const20, coeff4=>const29)
       port map( x=>s2-, y1=>s3-, y2=>s9+, y3=>s12+, y4=>s13+);

CM02p : CM generic map ( coeff1=>const02, coeff2=>const11,
          coeff3=>const25, coeff4=>const21, coeff5=>const30)
       port map( x=>s4+, y1=>s2-, y2=>s5-, y3=>s9-, y4=>s12-,
                                                      y5=>s13-);
CM02m : CM generic map ( coeff1=>const02, coeff2=>const11,
          coeff3=>const25, coeff4=>const21, coeff5=>const30)
       port map( x=>s4-, y1=>s2+, y2=>s5+, y3=>s9+, y4=>s12+,
                                                      y5=>s13+);

CM03p : CM generic map ( coeff1=>const07, coeff2=>const16,
          coeff3=>const26, coeff4=>const22, coeff5=>const31)
       port map( x=>s6+, y1=>s3+, y2=>s7-, y3=>s9+, y4=>s12-,
```

```
                                                        y5=>s13 -);
CM03m : CM generic map ( coeff1=>const07, coeff2=>const16,
            coeff3=>const26, coeff4=>const22, coeff5=>const31)
        port map( x=>s6 -, y1=>s3 -, y2=>s7 +, y3=>s9 -, y4=>s12 +,
                                                        y5=>s13 +);

CM04p : CM generic map ( coeff1=>const12, coeff2=>const27,
                            coeff3=>const23, coeff4=>const32)
        port map( x=>s8 +, y1=>s5 +, y2=>s9 -, y3=>s12 +, y4=>s13 +);
CM04m : CM generic map ( coeff1=>const12, coeff2=>const27,
                             coeff3=>const23, coeff4=>const32)
        port map( x=>s8 -, y1=>s5 -, y2=>s9 +, y3=>s12 -, y4=>s13 -);

CM05p : CM generic map ( coeff1=>const04, coeff2=>const09,
   coeff3=>const14, coeff4=>const18, coeff5=>const28, coeff6=>1)
     port map( x=>s10 +, y1=>s2 +, y2=>s3 +, y3=>s5 -, y4=>s7 +,
                                               y5=>s9 +, y6=>s15 -);
CM05m : CM generic map ( coeff1=>const04, coeff2=>const09,
   coeff3=>const14, coeff4=>const18, coeff5=>const28, coeff6=>1)
     port map( x=>s10 -, y1=>s2 -, y2=>s3 -, y3=>s5 +, y4=>s7 -,
                                               y5=>s9 -, y6=>s15 +);

CM06p : CM generic map ( coeff1=>const03, coeff2=>const08,
                 coeff3=>const13, coeff4=>const17, coeff5=>1)
        port map( x=>s12 +, y1=>s2 +, y2=>s3 +, y3=>s5 +, y4=>s7 -,
                                                        y5=>s16 -);
CM06m : CM generic map ( coeff1=>const03, coeff2=>const08,
                 coeff3=>const13, coeff4=>const17, coeff5=>1)
        port map( x=>s12 -, y1=>s2 -, y2=>s3 -, y3=>s5 -, y4=>s7 +,
                                                        y5=>s16 +);

CM07p : CM generic map ( coeff1=>const05, coeff2=>const10,
   coeff3=>const15, coeff4=>const19, coeff5=>const33, coeff6=>1)
     port map( x=>s14 +, y1=>s2 +, y2=>s3 +, y3=>s5 +, y4=>s7 -,
                                              y5=>s13 +, y6=>s17 -);
CM07m : CM generic map ( coeff1=>const05, coeff2=>const10,
   coeff3=>const15, coeff4=>const19, coeff5=>const33, coeff6=>1)
     port map( x=>s14 -, y1=>s2 -, y2=>s3 -, y3=>s5 -, y4=>s7 +,
                                              y5=>s13 -, y6=>s17 +);

end architecture SCHEMATIC;
```

The description of the SI system in the above file shows that it is made up of two basic cells: current mirrors and integrators (Hughes et al., 1996; Rudnicki et al., 2010), which are described in Sections 3.5 and 3.7. Further design stages depend on the choice of technology in which the integrated circuit is to be made. The integrator is a library cell in **SIstudio** and must be previously designed according to the indications in Rudnicki et al. (2010). Current mirrors, on the other hand, are parameterized cells and are automatically designed based on the generated grid. The basis for obtaining such grid are simulations of current mirrors in the **Eldo** or **Spice** environments for selected channel length ratios in the output stage of the mirror to the input stage. Sample results of these simulations for the *AMS* 0.6 μm technology are included in the file **spiceCM_AMS06U_w14.dat** as follows:

```
% Input data file for Current Mirror program
% (transistor length & width are in [um])

% AMS 0.6U L49 (am=4.0V)
% - min channel length: 0.6U, min channel width: 0.8U
% - grid: integral multiple of 0.05U
% - current mirror composed of nMOS transistors with
%   channel width w=1.4u
CURRENT MIRROR MAKER - INPUT DATA FILE

% Technology grid
GRID = 0.05

% NMOS transistors channel width
NMOS_WIDTH = 1.4

% transistor length to width (2D array),
% and bias current in [uA]:
% - results from diode connection simulations
LxW = 9
L = 0.6   1   2   3   4   5   6   7   8
W = 4.47   4.16   3.85   3.69   3.62   3.57   3.54   3.52   3.51
IB = 70   27   11   7   5   4   3.3   2.8   2.5

% Input/Output transistor (lin/lout) length
% to current gain (S)
S(LINxLOUT) = 9 x 8
LIN =  0.6 1 2 3 4 5 6 7 8
LOUT = 1 2 3 4 5 6 7 8

LIN_ARRAY = 0.6
S =  0.4577 0.2039 0.1301 0.0954 0.0753 0.0623 0.0530 0.0461

LIN_ARRAY = 1
S = 0.9685 0.4311 0.2748 0.2021 0.1591 0.1315 0.1122 0.0979

LIN_ARRAY = 2
S = 2.2231 0.9914 0.6316 0.4643 0.3661 0.3024 0.2577 0.2247

LIN_ARRAY = 3
S = 3.5053 1.5584 0.9939 0.7306 0.5757 0.4760 0.4047 0.3530

LIN_ARRAY = 4
```

```
S = 4.7658 2.1237 1.3545 0.9950 0.7843 0.6472 0.5528 0.4807

LIN_ARRAY = 5
S = 6.0715 2.7004 1.7218 1.2654 0.9979 0.8238 0.7028 0.6114

LIN_ARRAY = 6
S = 7.3254 3.2721 2.0874 1.5321 1.2082 0.9975 0.8483 0.7404

LIN_ARRAY = 7
S = 8.6135 3.8339 2.4457 1.7951 1.4157 1.1717 0.9970 0.8675

LIN_ARRAY = 8
S = 9.8951 4.4081 2.8097 2.0657 1.6298 1.3461 1.1488 1.0000
```

Simulations were carried out for nine transistor channel lengths given in μm in the line $L = 0.6, 1, \dots, 8$. The channel widths of the n-channel transistors are 1.4 μm. To obtain the appropriate channel widths of the p-channel transistors, a pair of complementary transistors in a diode connection is simulated. The results are contained in the next two lines. In the line $W = 4.47, 4.16, \dots, 3.52, 3.51$, the pMOS channel widths are given in μm for which the diode connection is balanced (voltage in the common node close to 0 V with symmetrical power supply), according to the adopted channel length. The next line shows the bias currents flowing through the diode connection of the transistors $IB = 70, 27, \dots, 2.8, 2.5$ in μA, also according to the channel length.

Current mirror simulations are performed for channel length ratios in all their combinations given in lines *LIN*, *LOUT*. The obtained values *S* of the scaling factors of the current mirrors are listed in the following lines. Based on these results, the **CMmaker** program generates the entire grid of scaling factors in the range of selected transistor channel lengths with a step equal to the technological pitch 0.05 μm. Missing points in relation to the simulation results are obtained by interpolation.

The program **CMmaker** calculates, based on the generated grid, the sizes of the transistor channels in the mirrors in the designed SI network. As an example, let's take the four-output current mirror *CM01p* in the network description. The scaling factors corresponding to the subsequent outputs are as follows: $const06 = 2.324$, $const24 = 0.536$, $const20 = 0.9375$, $const29 = 0.9826$. Figure 4.7 presents the window showing the next steps of the program **CMmaker**: 1 – loading the file with technological data, 2 – generating the grid, 3 – calculating the sizes of the current mirror transistor channels for the scaling factors entered in the table. The obtained results were saved in the following file:

```
CURRENT MIRROR - output data (Final parameters)
-----------------------------------------------------------
-----------------------------------------------------------
```

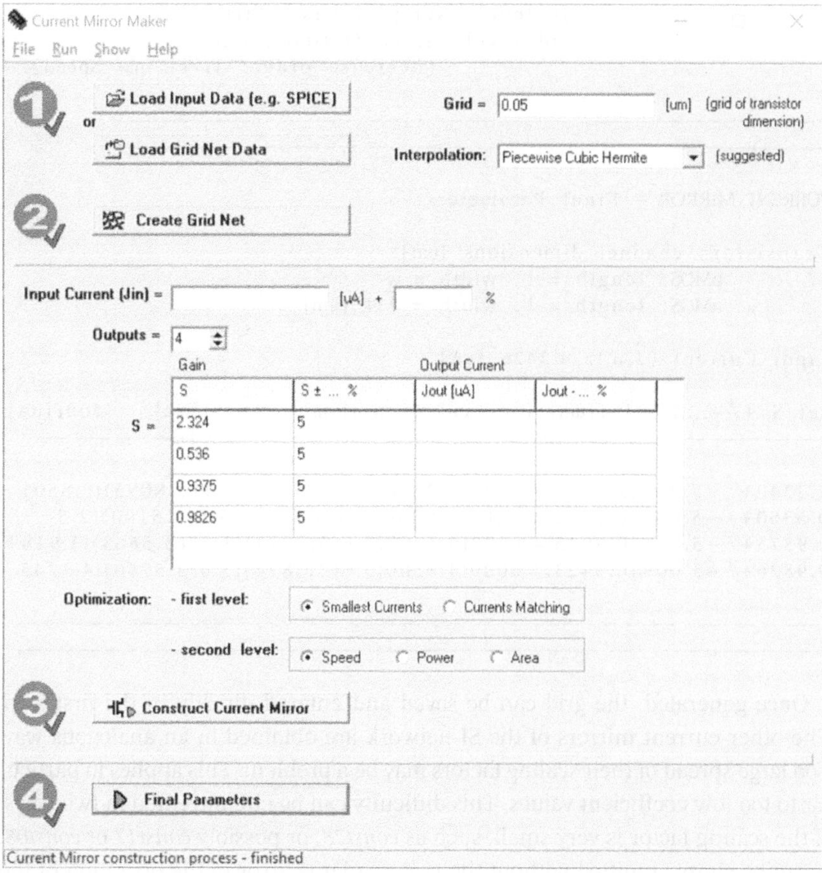

Figure 4.7 Window of **CMmaker** after the actions: loading of technology file, grid generation, and current mirror construction.

```
Input data file:
        C:\Andrzej\BOOK2\SIstudio\Siatka\spiceCM_AMS06U_w14.dat
Time-Date: 14:29:24 25.10.2022

Settings:
        Grid = 0.05 [um]
        Interpolation: Piecewise Cubic Hermite
        Input Current (Jin): not set
        Output Current (Jout):
                S = 2.3240 +/- 5.00%    |  not set
                S = 0.5360 +/- 5.00%    |  not set
                S = 0.9375 +/- 5.00%    |  not set
                S = 0.9826 +/- 5.00%    |  not set
        Optimization:
```

```
          – first level: scriptsizeest Currents
          – second level: Speed Optimization
                          (decrease order: first max Speed)

________________________________________________________________
________________________________________________________________

CURRENT MIRROR – Final Parameters:

Transistors channel dimensions [um]:
        pMOS: length = l, width = w
        nMOS: length = l, width = 1.4 [um]

Input Current (Jin): 4.3426 [uA]

set S +/–...%|S|lin[um]|lout[um]   |win[um]   |wout[um]    |Jout[uA]
____________________________________________________________________

2.3240+/–5.00%|2.2172|4.6000|2.2000|3.6(3.5874)|3.8(3.8095)|9.8501
0.5360+/–5.00%|0.5591|4.6000|8.0000|3.6(3.5874)|3.5(3.5100)|2.5
0.9375+/–5.00%|0.8935|4.6000|5.1000|3.6(3.5874)|3.55(3.5663)|3.9189
0.9826+/–5.00%|0.9423|4.6000|4.8500|3.6(3.5874)|3.6(3.5760)|4.1245

____________________________________________________________________
____________________________________________________________________
```

Once generated, the grid can be saved and entered directly in the first step. The other current mirrors of the SI network are obtained in an analogous way. Too large spread of their scaling factors may be a problem. This applies in particular to too low coefficient values. This difficulty can be circumvented in two ways. If the scaling factor is very small, such as *const28*, or possibly *const17* or *const08*, it can be simply omitted without the risk of a large error in the operation of the network. However, if the omission of a small factor would cause a significant error in signal processing, it would have to be implemented by cascading two, three or more current mirrors, each with a scaling factor equal to the square root, cube root, etc., of the single mirror factor, respectively.

The block diagram of the analysis filter is shown in Fig. 4.8. Basic blocks, integrators and current mirrors, and input and output signals are visible at a glance. However, the connections are difficult to trace in this diagram without referring to the text description in the file **SI1in3o.vhdl**. The input and output signals are, in a fully differential structure, the current equivalents of the differential voltage signals in the prototype gC circuit. In the current mirrors *CM05* and *CM06* the outputs, corresponding to the small scaling factors *const17* and *const28* mentioned earlier, have been omitted. To improve the common mode rejection ratio (CMRR), a compensation circuit (Szczesny, 2017) can be connected to the output of each integrator.

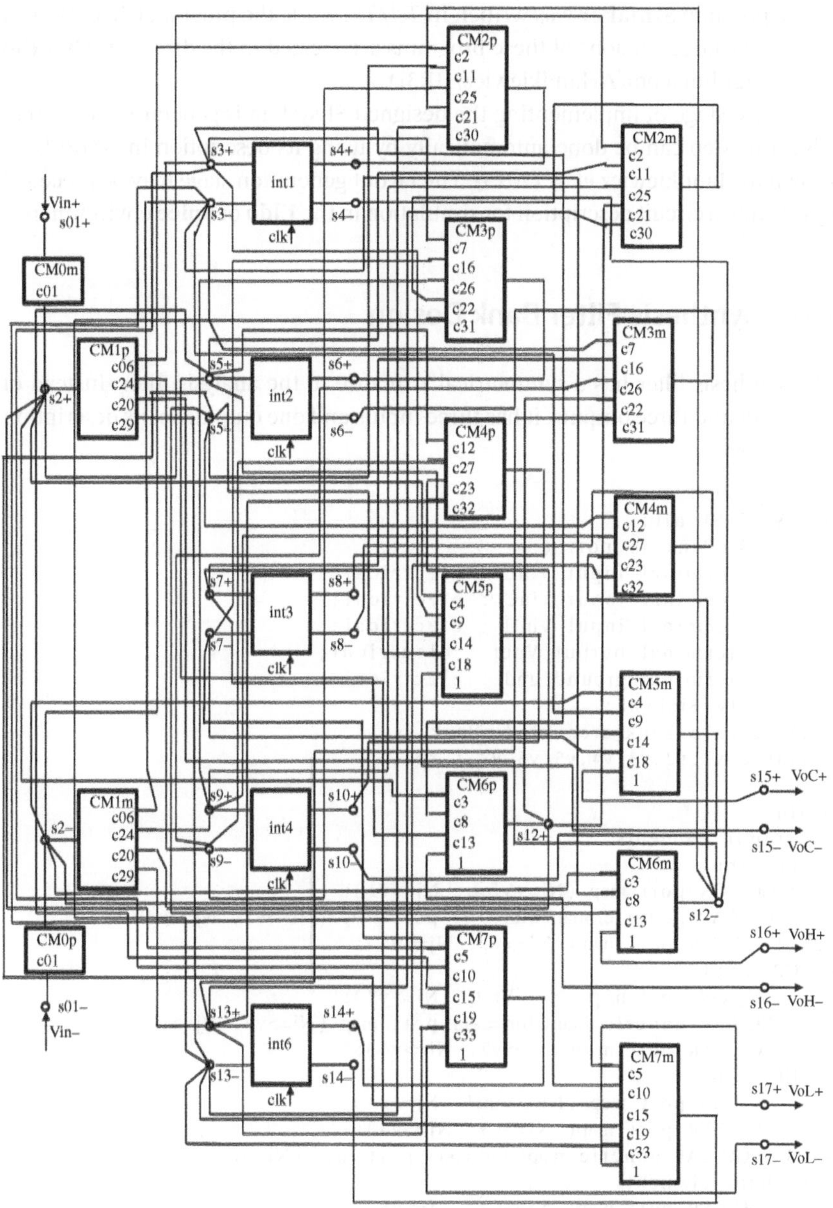

Figure 4.8 A block diagram of the analysis filter bank implemented in the SI technique.

The program **SImaker** was written in *Tcl/Tk*, while the program **CMmaker** in C^{++}. The source versions of these programs are located in the directory **SIstudio** at http://github.com/AHandkiewicz/IIR3D.

The last stage of implementing the designed SI system is generating its layout. Also, this step can be done automatically by using its description in AMPLE, as shown in Handkiewicz et al. (2015). The layout generation stage may be preceded by creating a circuit description for simulation in the **Eldo** or **Spice** environments.

4.6 Synthesis Filter Bank Design

The synthesis filter has a symmetrical structure to the analysis filter. Instead of one input and three outputs, it has three inputs and one output. So, its description is as follows:

```
library VLSI;
use VLSI.SI.all;
entity 3inBank is port (
        terminal input VinL : electrical;
        terminal input VinC : electrical;
        terminal input VinH : electrical;
        terminal output Vout : electrical;
        terminal ground gnd : electrical);
end entity 3inBank;
architecture bank_arch of 3inBank is
variable v1,v2,v3,v4,v5,v6,v7 : real;

begin
--- input elements
-- LP input
G1 : GC_CON port map (Ni=>VinL, Nj=>v1);
C1 : GC_CAP generic map(dim=>1) port map (Ni=>v1, Nj=>v4);
ig1 : GC_GYR port map( Ni=>v1, Nj=>v4);
-- CP input
G2 : GC_CON port map (Ni=>VinC, Nj=>v2);
C2 : GC_CAP generic map(dim=>1) port map (Ni=>v2, Nj=>v4);
ig2 : GC_GYR port map( Ni=>v2, Nj=>v4);
-- HP input
G3 : GC_CON port map (Ni=>VinH, Nj=>v3);
ig3 : GC_GYR port map( Ni=>v3, Nj=>v4);
--C3 : GC_CAP generic map(dim=>1) port map (Ni=>v3, Nj=>v4);
--- chain elements
ig4 : GC_GYR port map( Ni=>v4, Nj=>v5);
C4 : GC_CAP generic map(dim=>1) port map (Ni=>v4, Nj=>v5);
ig5 : GC_GYR port map( Ni=>v5, Nj=>v6);
C5 : GC_CAP generic map(dim=>1) port map (Ni=>v5, Nj=>v6);
ig6 : GC_GYR port map( Ni=>v6, Nj=>v7);
C6 : GC_CAP generic map(dim=>1) port map (Ni=>v6, Nj=>v7);
```

```
ig7 : GC_GYR port map( Ni=>v4, Nj=>v7);
ig8 : GC_GYR port map( Ni=>v5, Nj=>v7);
ig9 : GC_GYR port map( Ni=>v4, Nj=>v6);
-- output node
G4 : GC_CON port map (Ni=>v7, Nj=>gnd, Vo=>Vout);
--- additional capacitors
C7 : GC_CAP generic map(dim=>1) port map (Ni=>v4, Nj=>v6);
---C8 : GC_CAP generic map(dim=>1) port map (Ni=>v4, Nj=>v7);
C8 : GC_CAP generic map(dim=>1) port map (Ni=>v5, Nj=>v7);
--- gyrators from inputs to chain nodes
-- LP input
---ig11 : GC_GYR port map( Ni=>v1, Nj=>v5);
---ig12 : GC_GYR port map( Ni=>v1, Nj=>v6);
---ig13 : GC_GYR port map( Ni=>v1, Nj=>v7);
-- CP input
---ig21 : GC_GYR port map( Ni=>v2, Nj=>v5);
---ig22 : GC_GYR port map( Ni=>v2, Nj=>v6);
---ig23 : GC_GYR port map( Ni=>v2, Nj=>v7);
-- HP input
---ig31 : GC_GYR port map( Ni=>v3, Nj=>v5);
---ig32 : GC_GYR port map( Ni=>v3, Nj=>v6);
---ig33 : GC_GYR port map( Ni=>v3, Nj=>v7);
--- gyrators betwen inputs
ig41 : GC_GYR port map( Ni=>v1, Nj=>v2);
ig42 : GC_GYR port map( Ni=>v1, Nj=>v3);
ig43 : GC_GYR port map( Ni=>v2, Nj=>v3);

end architecture;
```

For the above architecture, a file **3in1o1D.sn** with structural numbers is generated in the **gCstudio** environment. The **t0** and initialization **init1D** files are the same as for the analysis filter. The file **t1** differs from **t0** in that the Kronecker delta is set not as the zero ($k = 0$) but as the fourth sample ($k = 4$). So, we will expect that the PR condition (4.5) will be satisfied with the scaling factor $c = 1$ and the delay $d = 4$. The most important difference compared to analysis filter design is in the additional digital networks mentioned in **note** in Section 1.1.2. They are now given by transfer function polynomials in the domain z designed in Section 4.4 of the analysis filter **zd1=zd2=zd3=z0, zn1=z1, zn2=z2, zn3=z3.**

After running **optimizer**

```
for ((i=0; i<=9; i++)); do
./optimizer -c 1 -x -n -t 0.01 -a t -i 1000 -d 16 1 1 3in1o1D.sn
                                                          init1D > "f$i"
    grep Goal "f$i"
    done
```

we get as one of the solutions:

```
0.924618422844948
0.002248882256470
```

```
1.454744483514349
0.108499056147961
0.145320707698078
1.293115593939024
-0.368206825512027
0.829217114171472
-0.735552353315427
0.862568144402959
-0.573211301005377
1.116953501466046
0.990811204951248
0.693160662147604
-0.126446396123436
0.236072322949239
0.216512254585322
1.469862330261862
0.380703795807673
0.691282947928335
1.499990131070924
1.499979888530192
0.915914176997341
0.091041457784236
```

for the objective function with a value equal to 0.000001121022063. Fig. 4.9 illustrates the effect of the designed synthesis filter operation. The samples marked with crosses mean the output signal $y(k)$, $k = 0, 1, 2, \ldots, 15$, of the FI ADC in the absence of the synthesis filter, while the samples marked with circles show that condition (4.5) is, for the designed filter, fully achieved.

The digital synthesis filter, which is the equivalent of the prototype gC circuit, is obtained on the basis of its transfer function in the z domain. The polynomial coefficients of these transfer functions are stored in the files

z0:

1	
5	
0	9.66298485
1	15.70791245
2	7.73829556
3	2.26185966
4	0.76396036
5	0.16805857

z1:

1	
5	
0	−0.91784942
1	−0.74894792
2	8.28209972
3	18.04695702

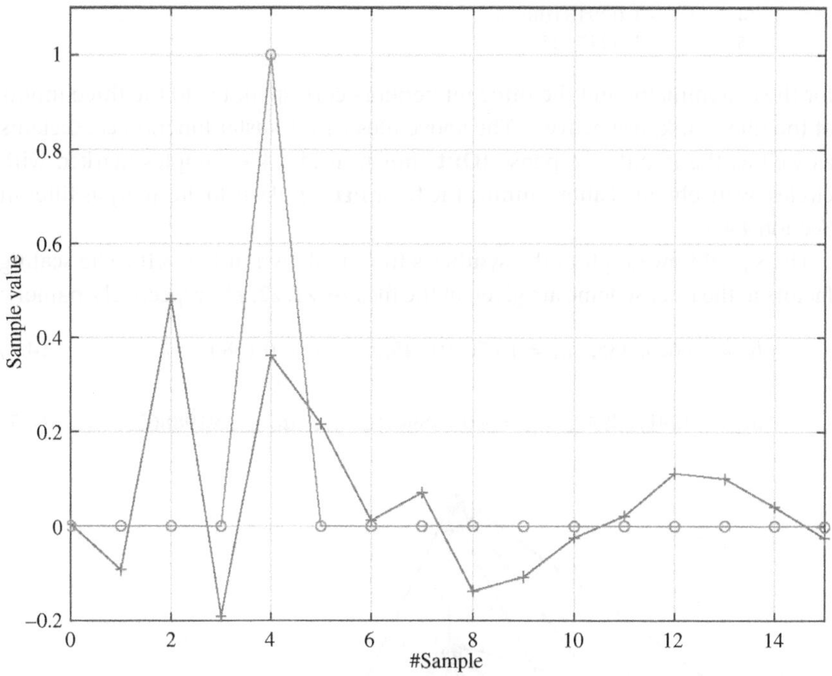

Figure 4.9 Perfect reconstruction achieved with the use of the designed synthesis filter (samples marked with circles) compared to FI ADC response (samples marked with crosses) when this filter is not present.

4	12.85364532
5	2.91988635

z2:

1	
5	
0	4.08900928
1	1.01439738
2	−10.43551636
3	−9.91287041
4	−3.29167557
5	−0.73970926

z3:

1	
5	
0	−2.06805563
1	−4.31705809
2	3.75784039
3	5.97148132

```
4        -1.02915108
5        -1.02471256
```

for the denominator and the three numerators corresponding to the three inputs of the filter bank, respectively. The above files with transfer function coefficients as well as the impulse response **tOut** shown in Fig. 4.9 (samples marked with circles) were obtained after running the filter **Fil**, similarly to the analysis filter in Section 4.4.

The signal flow graph of the synthesis filter is shown in Fig. 4.10. The scaling factors in the filter scheme are given in the files **z0, z1, z2, z3**, respectively, namely:

$$b_0 = 9.66298485, \quad b_1 = 15.70791245, \ldots, \quad b_5 = 0.16805857, \tag{4.6}$$

$$a_{l0} = -0.91784942, \quad a_{l1} = -0.74894792, \ldots, \quad a_{l5} = 2.91988635, \tag{4.7}$$

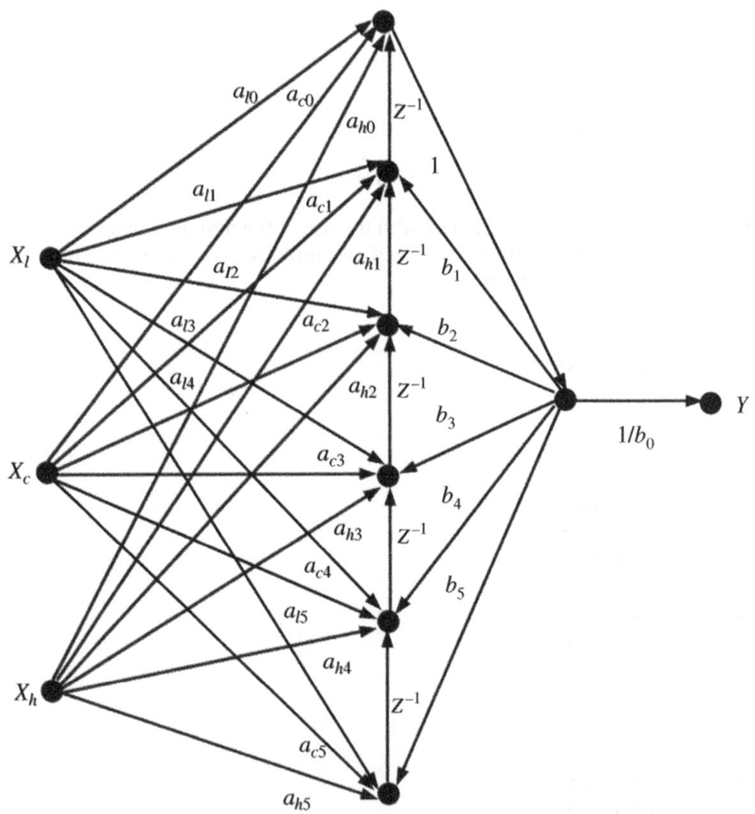

Figure 4.10 Digital implementation of the three-input, one-output synthesis filter.

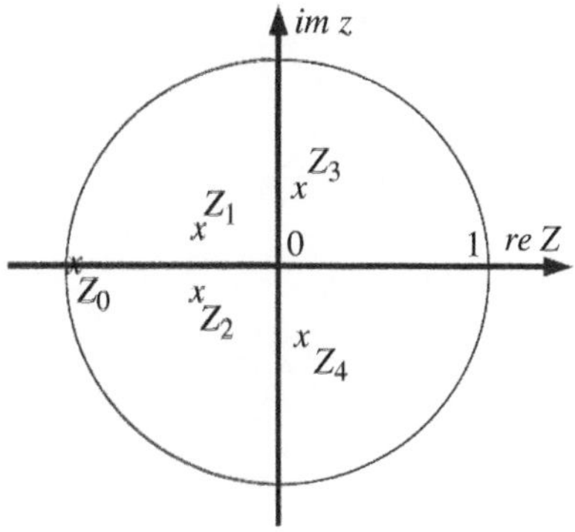

Figure 4.11 Root locations of the stable IIR synthesis filter.

$$a_{c0} = 4.08900928,\ \ a_{c1} = 1.01439738, \ldots,\ \ a_{c5} = -0.73970926, \tag{4.8}$$

$$a_{h0} = -2.06805563,\ \ a_{h1} = -4.31705809, \ldots,\ \ a_{h5} = -1.02471256. \tag{4.9}$$

The spread of values of the above coefficients is $a_{l3}/b_5 = 18.04695702/0.16805857 = 107.38$. The received circuit has the real pole $z_0 = -0.99430$ and two pairs of coupled complex poles $z_{1/2} = -0.37925 \pm 0.13103i$, $z_{3/4} = 0.06362 \pm 0.32341i$ whose position, presented in Fig. 4.11, confirms stability of the synthesis filter. This stability property is inherited from the prototype passive gC circuit.

5
Digital Implementation of Multidimensional Networks

The methods presented in Sections 2.2 and 2.3 apply to gC circuits with any number of ports and any dimension. In this chapter, we will present an example of using the three-dimensional (3D) gC circuit for video filtering. The implementation of this filter in an field programmable gate array (FPGA) matrix demonstrates the possibility of real-time 3-D video processing. The use of such a filter to generate subpixels and intermediate pixels related to the calculation of the motion vector (MV) is also presented. Note that due to the lack of appropriate methods for designing 3D infinite impulse response (IIR) filters, currently available video sequence processing standards only use one-dimensional finite impulse response (FIR) filters.

As an introduction to video filtration, we will present issues related to compression standards at the beginning of the chapter. The division of a video frame into blocks is of fundamental importance in standards. In Section 5.1.1, this issue is illustrated with the AV1 codec. The role of transform coding is illustrated with a simple example of discrete cosine transform (DCT). The idea of entropy coding in the standards is also illustrated with a simple example of arithmetic coding.

5.1 Basics of Compression Standards

Signal compression methods, especially video, are subject to continuous development related to the advancement of hardware feasibility. Processors, memories, and FPGAs operating in real time enable the implementation of algorithms of increasing computational complexity. At the same time, the compression rate and image quality increase. Various compression formats, initially associated mainly with TV signals via satellite or cable transmission, are now strongly connected with social media. The small silicon area and reduced power consumption of digital circuits enable the construction of codecs embedded in portable devices. The rapid development of compression methods is also due to the fact that open access

Methods and Tools for Computer-aided Design of Multidimensional Filters in Signal Processing,
First Edition. Andrzej Handkiewicz.

to the offered compression standards is possible. The basis of all compression standards is dividing video frames into blocks and transmitting encoded blocks instead of individual pixels. An example is the AV1 format, developed since 2018 and made available by the Alliance for Open Media consortium formed in 2015 (de Rivaz and Haughton, 2022). In Sections 5.1.1 and 5.1.2, we will discuss the basic operations related to the compression of video signals.

Color images and video can be represented in different color spaces. For images generated in image sensing arrays or displayed on screens, red (R), green (G), and blue (B) components are used. Image sensing arrays are built with the standard CMOS process and are composed of photodiodes or phototransistors. To obtain color images, the sensors of the array are alternately covered by red, green, and blue (RGB) filters following the Bayer pattern, as shown in Fig. 5.1 for an 8×8 pixel block. However, there is a strong correlation between these components in natural images. In image processing, it is advisable to use a color space with luminance (luma) (Y) and chrominance (chroma) C_B and C_R components. Different transformations between these spaces exist. One of them is defined as

$$\begin{bmatrix} Y \\ C_B \\ C_R \end{bmatrix} = \frac{1}{2} \begin{bmatrix} 0.598 & 1.174 & 0.228 \\ -0.338 & -0.662 & 1.000 \\ 1.000 & -0.838 & 0.162 \end{bmatrix} \cdot \begin{bmatrix} R \\ G \\ B \end{bmatrix}. \tag{5.1}$$

The much greater sensitivity of the human visual system to the luma than to the chroma signals enables compression by removing pixels from the chroma components. For example, 4 : 2 : 0 compression means leaving only 1 pixel of both chroma components for every group of 4 pixels of luminance.

5.1.1 Block Partition

Processing of individual images or their sequences is most often performed in blocks: two-dimensional (2D), briefly called blocks in the case of images or video frames, or 3D blocks in the case of video.

Division into blocks in modern standards is a multilevel operation. In the AV1 format, an initial block of size 128×128, called a superblock, may be subject to

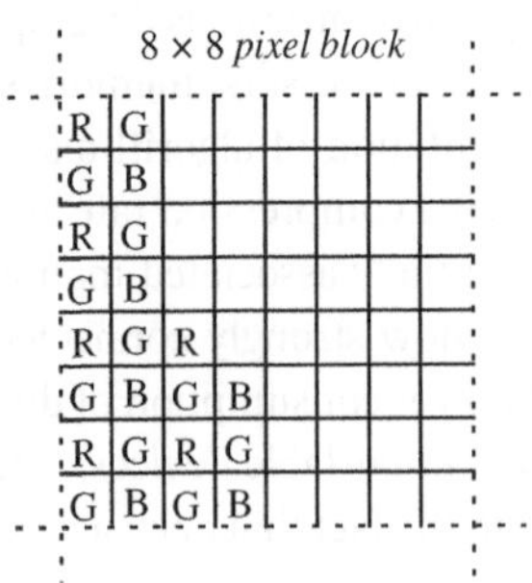

Figure 5.1 Pixel block covered by red, green, and blue (RGB) filters, following the Bayer pattern.

subsequent splits presented in a tree as a partitioning structure. Each node in the tree may terminate in a square block smaller than the superblock. This operation is called NONE. A square block, you can recursively (R) divide it into subsequent square blocks, but not smaller than 4 × 4. This operation is called SPLIT (Bender et al., 2023). The final splits into two, three, or four blocks are as follows:

- HORZ or VERT: the block is split into two rectangles in a horizontal or a vertical position,
- HORZ_A, HORZ_B, VERT_A, VERT_A: the block is split into three blocks – two squares and a one rectangle,
- HORZ_4 or VERT_4: the block is split into four rectangles in a horizontal or a vertical position.

The above kinds of splits are shown in Fig. 5.2. Such block partition is implemented in AV1 for prediction. Additional partition of squares into four blocks or rectangles into two blocks can be allowed for a transform coding described in Section 5.1.2.

An example of the block partition is shown in Figs. 5.3 and 5.4. The first figure shows the partition tree. Leaves or nodes marked with squares indicate prediction blocks and marked with circles indicate transform blocks. Note that the example illustrates the use of the additional option in the AV1 format of dividing a rectangular block into two transform blocks. The criterion for ending the process of dividing into transform blocks is low variability of sample values in the block. As will be explained in Section 5.1.2, meeting such a condition will allow, after applying transform coding, to significantly reduce the number of samples to be quantized in the block. Figure 5.4 shows the division into blocks corresponding to the tree in Fig. 5.3. Prediction blocks are marked with solid lines, and transform blocks with dashed lines.

5.1.2 Motion Compensation, Transform Coding, Quantization, and Entropy Coding

Motion compensation is based on the assumption that a block similar to the one encoded in the current frame has previously occurred. If the similar block is found in the current frame then the process is called intraframe encoding and if in a previously encoded frame, used as the reference frame, then it is called interframe encoding. Assuming that the block is moving between frames, a motion vector (MV) is determined following block matching, called motion estimation. Compression then involves sending the encoded difference between both blocks and the MV.

The optimum transform for image compression is the Karhunen–Loeve transformation (KLT) (Jain, 1989) because it packs the greatest amount of energy in

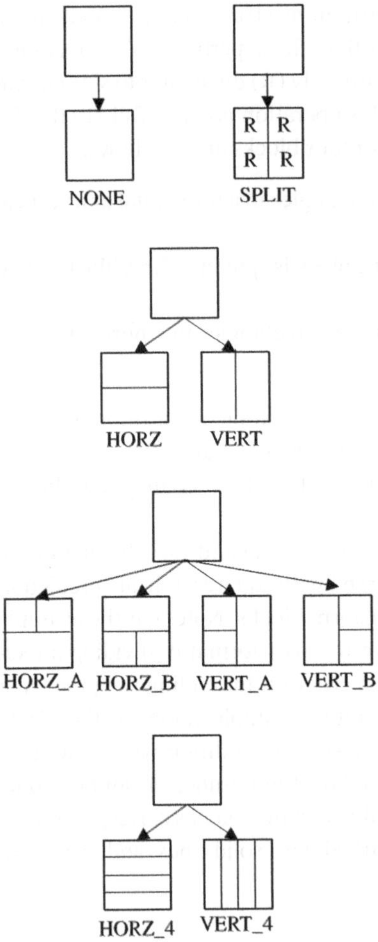

Figure 5.2 AV1 block partition for prediction and transformation. The block R can be recursively split into subsequent square blocks.

the smallest number of elements in the frequency domain of a 2D signal and minimizes the total entropy of the signal sequence. Unfortunately, the basis functions of KLTs are image dependent, which is the most important implementation-related deficiency. It is observed that the 2D DCT has the output close to the output produced by the KLT (Bhaskaran and Konstantinides, 1995), and uses image-independent basis functions. Hence, DCT-based image coding is applied in all video compression standards. Other types of transforms available for example in AV1 are: Asymmetrical Discrete Sine Transform (ADST), flipADST, and Identity Transform (IDTX).

As it was shown in Section 5.1.1, the transform blocks can be in different sizes. We will illustrate DCT operation assuming that the image is divided into 8×8

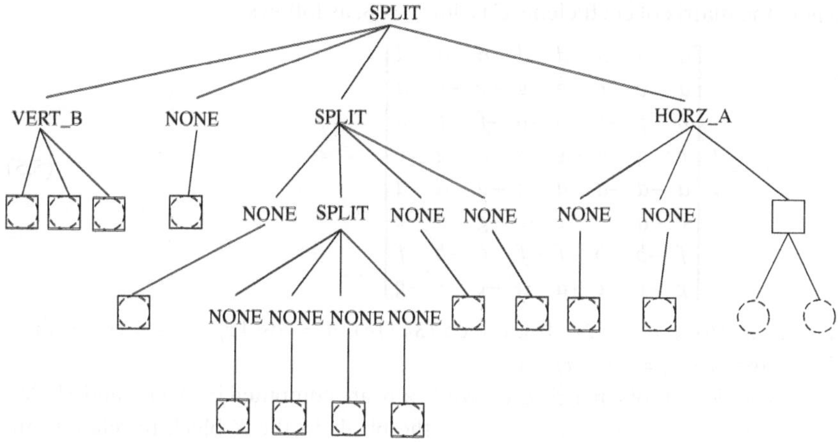

Figure 5.3 Block partitioning structure as a tree.

Figure 5.4 Block partition into subblocks for prediction (solid line) and for transform (dashed line).

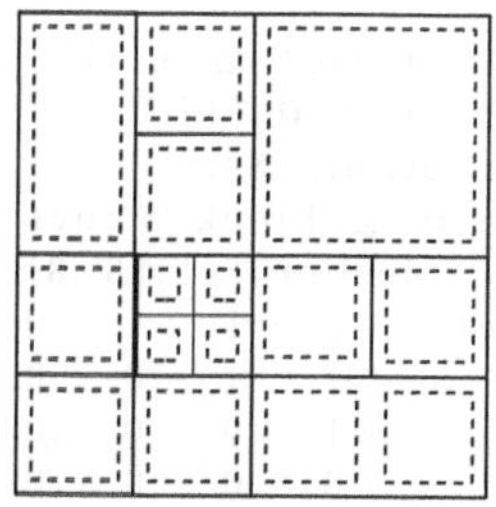

blocks in the spatial domain, and DCT transforms them into 8×8 blocks in the 2D frequency domain. Such block size is convenient with respect to computational complexity.

The 2D DCT can be expressed as

$$y_{kl} = \frac{c(k)c(l)}{4} \sum_{i=0}^{N-1} \sum_{j=0}^{N-1} x_{ij} \cos \frac{(2i+1)k\pi}{2N} \cos \frac{(2j+1)l\pi}{2N}, \tag{5.2}$$

where $k, l = 0, 1, \ldots, N-1$ and

$$c(k) = \begin{cases} \frac{1}{\sqrt{2}}, & k = 0, \\ 1, & k \neq 0. \end{cases} \tag{5.3}$$

Assuming that the matrices Y and X are composed of elements y_{ij} and x_{ij}, $i, j = 0, 1, \ldots, N-1$, respectively, relation (5.2) can also be written in the matrix form as

$$Y = CXC', \tag{5.4}$$

where the matrix of coefficients C is, for $N = 8$, as follows

$$C = \frac{1}{2}\begin{bmatrix} d & d & d & d & d & d & d & d \\ a & c & e & g & -g & -e & -c & -a \\ b & f & -f & -b & -b & -f & f & b \\ c & -g & -a & -e & e & a & g & -c \\ d & -d & -d & d & d & -d & -d & d \\ e & -a & g & c & -c & -g & a & -e \\ f & -b & b & -f & -f & b & -b & f \\ g & -e & c & -a & a & -c & e & -g \end{bmatrix}, \tag{5.5}$$

$a = \cos(\pi/16)$, $b = \cos(2\pi/16)$, $c = \cos(3\pi/16)$, $d = \cos(4\pi/16)$, $e = \cos(5\pi/16)$, $f = \cos(6\pi/16)$, $g = \cos(7\pi/16)$.

The coefficients of a matrix C of size $N \times N$ are computed in Octave and Matlab by the function $C = dctmtx(N)$. While the pixels in the X block in relationship (5.4) are distributed evenly, in the block Y they are concentrated in its upper left corner. We will illustrate this property with the following example in the Octave environment:

```
I = imread('moon.tif ');
I = im2double(I);
C = dctmtx(8);
dct = @(block_struct) C * block_struct * C';
B = blockproc(I,[8 8],dct);
mask = [1   1   1   0   0   0   0   0
        1   1   0   0   0   0   0   0
        1   0   0   0   0   0   0   0
        0   0   0   0   0   0   0   0
        0   0   0   0   0   0   0   0
        0   0   0   0   0   0   0   0
        0   0   0   0   0   0   0   0
        0   0   0   0   0   0   0   0];
B2 = blockproc(B,[8 8],@(bs) mask.* bs);
invdct = @(block_struct) C' * block_struct * C;
I2 = blockproc(B2,[8 8],invdct);
imshow(I), figure , imshow(I2)
```

In this example, DCT (*dct*) and inverse DCT (*invdct*) are calculated according to formula (5.4). Performing these operations on blocks of a selected frame is possible in Octave using the *blockproc* function. The inverse DCT is calculated not for all 8×8 pixels of the block but only for the six indicated by 1's in the masking matrix *mask*. The compression efficiency is therefore greater than 0.1. The result of such compression is shown in Fig. 5.5. The original image is on the left and the compressed image is on the right of the figure.

The DCT coefficients are scanned in a zig-zag order, as shown in Table 5.1. This is due to the shown DCT property of locating significant pixel values in the upper left corner of the block. Selected samples, for example $0, 1, 2, 3, 4, 5$ are quantized. In

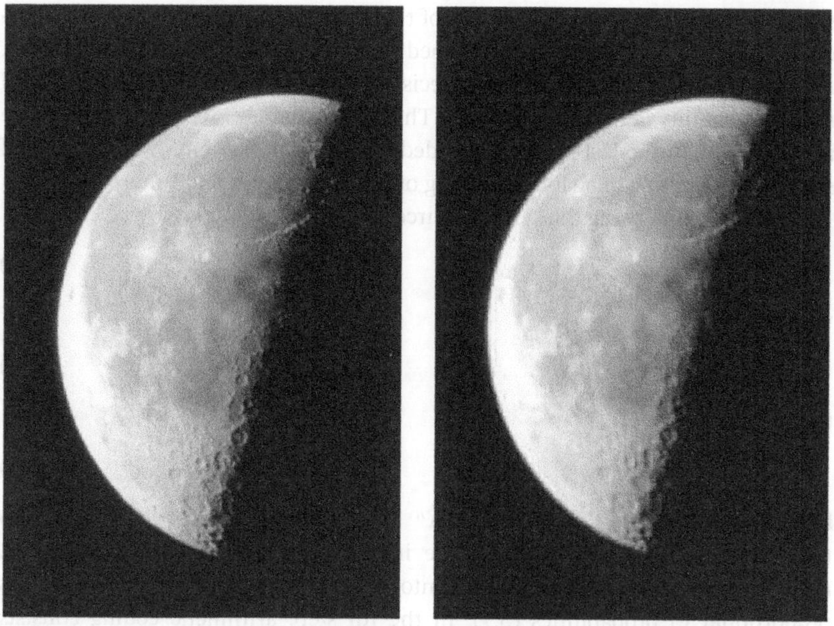

Figure 5.5 Image compression with the use of DCT. The original frame is on the left and the compressed frame is on the right of the figure.

Table 5.1 Zig-zag scan order.

0	1	5	6	14	15	27	28
2	4	7	13	16	26	29	42
3	8	12	17	25	30	41	43
9	11	18	24	31	40	44	53
10	19	23	32	39	45	52	54
20	22	33	38	46	51	55	60
21	34	37	47	50	56	59	61
35	36	48	49	57	58	62	63

the AV1 standard, 8-, 10-, and 12-bit pixel representations are possible. The resulting sequence of numbers is then entropy-coded using arithmetic coding.

Arithmetic coding involves assigning, in successive steps, intervals with lengths proportional to probabilities of source symbols selected to encode successive bits of a word. Since the sum of the symbol probabilities in the source is equal to 1, the initial interval has length equal to 1. Subsequent subintervals are shorter and shorter, which means that the length of the encoded word is limited by the precision of arithmetic operations performed in the hardware. The encoder adopted, for example in AV1 allows for 15-bit precision. The last interval has a length equal to the probability of the encoded word. The code sequence is created based on the average value in the last interval recorded in the binary system. A description of such an algorithm can be found, among others, in Howard and Vitter (1994).

We assume that using the symbol source

$$X = \{x_1, x_2, \dots, x_j, \dots, x_k\}, \tag{5.6}$$

with probabilities

$$P(x_1),\ P(x_2), \dots, P(x_j), \dots, P(x_k), \tag{5.7}$$

the word

$$a = (a_1, a_2, \dots, a_i, \dots, a_n), \tag{5.8}$$

belonging to the nth extension $A = X^n$ of the source X has been created. So the encoding is done in n steps. The interval obtained in the $(i-1)$th step, $i = 1, 2, \dots, n$, can always be divided into a set of subintervals with the lengths proportional to probabilities (5.7). In the ith step, arithmetic coding consists of selecting the jth subinterval for which $x_j = a_i$. We will mark this choice symbolically

$$I_i = \{I_{i-1}\}|_j, \tag{5.9}$$

where the length l_i of the subinterval I_i is:

$$l_i = P(x_j) l_{i-1}. \tag{5.10}$$

The selected subinterval starts at the point

$$s_i = s_{i-1} + l_{i-1} \sum_{p=1}^{j-1} P(x_p), \tag{5.11}$$

where $\sum_{p=1}^{j-1} P(x_p)$ is the cumulative probability of 0 value for $j \leq 1$. The end of the interval is determined by its starting value (5.11) plus its length (5.10).

Relationship (5.10) shows that the interval in the last step $i = n$ has the length l_n equal to the probability of word (5.8):

$$l_n = P(a_1)P(a_2) \dots P(a_n) = P(a). \tag{5.12}$$

For this interval, the average value M in a binary notation is calculated. As the final codeword, the most significant m bits of M are taken, where:

$$m = \left\lfloor \log_2 \frac{1}{P(a)} \right\rfloor + 1. \tag{5.13}$$

The decoding procedure is performed based on the estimated average $\widehat{M}$ calculated for m bits of the codeword. The procedure consists in finding the subinterval I_i to which the estimated average is included:

$$\widehat{M} \epsilon I_i = \{I_{i-1}\}|_j, \tag{5.14}$$

i.e. a j ($j = 1, \dots, k$) is found for which

$$s_i = s_{i-1} + l_{i-1} \sum_{p=1}^{j-1} P(x_p) \leq \widehat{M} < s_i + P(x_j) l_{i-1} = s_i + l_i. \tag{5.15}$$

For $j = k$ the right inequality < of the above relation can be replaced by ≤, as will be done in an example.

We will illustrate the above relationships with simple examples.

Example 5.1 *Let us assume that the symbol source has two elements,* $X = \{x_1, x_2\} = \{0, 1\}$, *with probabilities* $P(0) = 1/4$, $P(1) = 3/4$ *and we encode its fifth extension* X^5. *As an example, let's choose from this extension the word* $a = (1, 0, 1, 1, 0)$, *whose probability is* $P(a) = P(1)^3 P(0)^2 = 27/1024$. *The number of codeword bits after arithmetic encoding, according to (5.13), is then:*

$$\lceil 5.245 \dots \rceil + 1 = 7. \tag{5.16}$$

The use of equations (5.10) and (5.11) is shown in Table 5.2. The initial values are: $I_0 = [0, 1]$, $l_0 = 1$, *and* $s_0 = 0$. *The average value in the last interval* I_5 *is therefore as follows:*

$$M = s_5 + \frac{l_5}{2} = \frac{707}{2048} = 0.3452148 \dots_{(10)} = 0.01011000 \dots_{(2)}. \tag{5.17}$$

The seven most significant bits in (5.17) give the codeword $a_c = (0, 1, 0, 1, 1, 0, 0)$ *for the selected* $a = (1, 0, 1, 1, 0)$.

Table 5.2 Arithmetic encoding of the word $a = (1, 0, 1, 1, 0)$.

i	$a_i = x_j$	j	l_i (5.10)	s_i (5.11)	$I_i = \{I_{i-1}\}\|_j$
1	$a_1 = x_2 = 1$	2	$\frac{3}{4} \cdot 1 = \frac{3}{4}$	$0 + 1 \cdot \frac{1}{4} = \frac{1}{4}$	$\left[\frac{1}{4}, 1\right]$
2	$a_2 = x_1 = 0$	1	$\frac{1}{4} \cdot \frac{3}{4} = \frac{3}{16}$	$\frac{1}{4} + 0 = \frac{1}{4}$	$\left[\frac{1}{4}, \frac{7}{16}\right)$
3	$a_3 = x_2 = 1$	2	$\frac{3}{4} \cdot \frac{3}{16} = \frac{9}{64}$	$\frac{1}{4} + \frac{3}{16} \cdot \frac{1}{4} = \frac{19}{64}$	$\left[\frac{19}{64}, \frac{7}{16}\right]$
4	$a_4 = x_2 = 1$	2	$\frac{3}{4} \cdot \frac{9}{64} = \frac{27}{256}$	$\frac{19}{64} + \frac{9}{64} \cdot \frac{1}{4} = \frac{85}{256}$	$\left[\frac{85}{256}, \frac{7}{16}\right]$
5	$a_5 = x_1 = 0$	1	$\frac{1}{4} \cdot \frac{27}{256} = \frac{27}{1024}$	$\frac{85}{256} + 0 = \frac{85}{256}$	$\left[\frac{85}{256}, \frac{367}{1024}\right)$

Table 5.3 Example of arithmetic decoding.

i	s_i (5.15)	l_i (5.15)	$I_i = \{I_{i-1}\}I_j$ (5.14)	j	$a_i = x_j$
1	$\frac{1}{4}$	$\frac{3}{4}$	$\left[\frac{1}{4}, 1\right]$	2	$a_1 = x_2 = 1$
2	$\frac{1}{4}$	$\frac{3}{16}$	$\left[\frac{1}{4}, \frac{7}{16}\right)$	1	$a_2 = x_1 = 0$
3	$\frac{19}{64}$	$\frac{9}{64}$	$\left[\frac{19}{64}, \frac{7}{16}\right]$	2	$a_3 = x_2 = 1$
4	$\frac{85}{256}$	$\frac{27}{256}$	$\left[\frac{85}{256}, \frac{7}{16}\right]$	2	$a_4 = x_2 = 1$
5	$\frac{85}{256}$	$\frac{27}{1024}$	$\left[\frac{85}{256}, \frac{367}{1024}\right)$	1	$a_5 = x_1 = 0$

Let us note here that a significant increase in coding efficiency can only be achieved for very long codewords, which requires high computing power.

Example 5.2 *For the codeword* $a_c = (0, 1, 0, 1, 1, 0, 0)$ *the decoder calculates the estimated average which in our example is* $\widehat{M} = \frac{1}{4} + \frac{1}{16} + \frac{1}{32} = \frac{11}{32}$. *In the successive steps,* $i = 1, \ldots, n$ *the jth source symbol is found according to relation (5.14). The values of* s_i *and* l_i *which fulfil relation (5.15) are shown in the 2nd and 3rd columns of Table 5.3. In the 4th column, the intervals into which the estimated average* $\widehat{M}$ *belongs are presented. The last column of the table contains the decoded word* $a = (1, 0, 1, 1, 0)$.

5.2 Three-Dimensional IIR Filters for Videos

Due to the dominating use of digital circuits in signal processing, the description of filters in the z domain is the most useful. There are two types of such filters: FIR and IIR (Lyons, 2001; Oppenheim et al., 1999). The first type of filter has the polynomial of the transfer function denominator simply equal to 1. FIR filters are obtained directly in the z domain. For IIR filters such direct methods do not exist and they must be designed as equivalents of prototype analog circuits described in the s domain. Presented in Chapter 1, RLC ladder structures, which include Butterworth, Chebyshev, and elliptic approximations of filter characteristics, are commonly used to implement IIR filters. Their transfer functions are obtained in the z domain using bilinear transformation (2.3) between the s and z domains. These methods apply only to two-port networks whose transfer functions are described by polynomials of only one complex variable s. To obtain a more general approach, multiport and multidimensional gyrator–capacitor (gC) circuits can be implemented, which is possible with the use of the tools presented in Chapter 2.

In this section, we will show the design of 3-D IIR filters for videos. Gyrator–capacitor circuits are the prototype circuits of such filters and enable their digital implementation using the bilinear transformation. Filtering of videos occurs at many stages of their processing, in particular those related to compression. However, the one-dimensional smooth, regular, and sharp filters used for this purpose have a FIR (Han et al., 2021). This is due to the lack of methods for designing stable multidimensional IIR digital filters. The 2-D IIR filter proposed in Tu and Chien (2021), for example has limited bounded input/bounded output stability only for signals and filter coefficients in the range from 0 to 1.

The gC prototype circuit, presented in this section, is implemented as a digital IIR filter of a video divided into 3D blocks with dimensions $d \times d \times d$ pixels, where $d = 16$. The first two edges of the 3D block correspond to the rows and columns in the frames and the third edge corresponds to the particular frames. The filter was implemented in the C++ language, using the **SNc** and **MDp** classes presented in Chapter 2. The source **filter2.cpp** and compiled **Fi2** versions of this filter are in the *Filter* directory, as are **filterl.cpp** and **Fil** from Section 4.3. The filter in the **filter2.cpp** version differs from the **filterl.cpp** version in bidirectional processing. The filtration is performed on the samples from the beginning of the sequence to its end and then from the last to the first sample. Filtering in different directions is typical in image processing (Bu et al., 2021; Tu and Chien, 2021).

The division into blocks was made using the **VideoReader** and **VideoWriter** functions available in the Octave environment. Two custom functions have been developed for this purpose:

videoCap – function captures blocks tIij, $i = 1, \dots, Nr$, $j = 1, \dots, Nc$ from a given sequence, which is then subject to 3-D filtering,

videoInt – function integrates the filtered blocks tOutij, $i = 1, \dots, Nr$, $j = 1, \dots, Nc$ into a video, visualizing the filtering effect,

where i and j are the row and column numbers of the blocks. The number of blocks ($Nr \times Nc$) depends on the number of rows (nr) and columns (nc) in the video frame and of the block size.

A common artifact when processing images divided into blocks is that the edges of the blocks are visible in frames (Chen et al., 2001; Maggioni et al., 2012; Norkin et al., 2012). In terms of filtration, this phenomenon is associated with initial conditions that require the use of prior samples, of the number n equal to the order of the filter. To avoid block edges, in the presented filtering method an n-pixel-thick border is created around all blocks in a frame. The border in the ij block coincides with the pixels on the edges of the adjacent blocks $i(j-1), (i-1)j, i(j+1), (i+1)j$. The number of blocks in rows and columns is, therefore:

$$Nr = \lfloor (nr - 2n)/(d - 2n) \rfloor, \quad Nc = \lfloor (nc - 2n)/(d - 2n) \rfloor. \tag{5.18}$$

Filtering of each 3D block is done with the use **FI2** in two directions: in the direction of moving frames, and then in the opposite direction. The division of the filtered video into 3D blocks, or their reconstruction, is done using the functions presented above. Since the filter is second order in each dimension, all 3D blocks that make up the video sequence contain a 2-pixel-thick border around each block, which is discarded when the blocks are re-sequenced, and the entire video is made up of 3D blocks containing $12 \times 12 \times 16$ pixels.

The prototype gyrator–capacitor filter will be, as in Section 4.3, also designed by comparing the time responses actual versus expected. For the 3D filters discussed in this section, such signals are artificial sequences generated in the Octave environment using the following *videoS* function:

```
%artificial videos
pkg load video
clear all;
d=6; %white area dimension: dxd
rows=16;
cols=16;
fras=16;
val = 1;
%zeroing of matrices
for i=1:rows
  for j=1:cols
   for k=1:fras
    I(i,j,k)=0;
        Is(i,j,k)=0;
   end
  end
end
%sharp image
for k=1:fras
        for k1=0 : d+1
         for k2=0 : d+1
      if 1 < k+k1 && 1 < k+k2 && k+k1 <= rows+1 && k+k2 <= cols+1
           Is(k+k1-1,k+k2-1,k)=val;
          end
         end
        end
end
%smooth image
dd=d+2;
for k=1:fras
        for k1=0 : dd+1
         for k2=0 : dd+1
```

```
        if 2 < k+k1 && 2 < k+k2 && k+k1 <= rows+2 && k+k2 <= cols+2
             I(k+k1-2,k+k2-2,k)=val/3;
            end
           end
          end
end
dd=d;
for k=1:fras
          for k1=0 : dd+1
           for k2=0 : dd+1
        if 1 < k+k1 && 1 < k+k2 && k+k1 <= rows+1 && k+k2 <= cols+1
             I(k+k1-1,k+k2-1,k)=2*val/3;
            end
           end
          end
end
dd=d-2;
for k=1:fras
          for k1=0 : dd+1
           for k2=0 : dd+1
        if 1 <= k+k1 && 1 <= k+k2 && k+k1 <= rows && k+k2 <= cols
             I(k+k1,k+k2,k)=val;
            end
           end
          end
end

w = VideoWriter ("t0.avi");
 open (w);
   for k=1:fras
     for j=1:cols
      for i=1:rows
       z(i,j,1)=z(i,j,2)=z(i,j,3)=Is(i,j,k);
      end
     end
    image (z);
%       axis off;
    drawnow
    writeVideo (w, getframe (gcf));
   end
 close (w)

w = VideoWriter ("t1.avi");
 open (w);
   for k=1:fras
     for j=1:cols
      for i=1:rows
       z(i,j,1)=z(i,j,2)=z(i,j,3)=I(i,j,k);
      end
     end
    image (z);
```

```
%         axis off;
      drawnow
      writeVideo (w, getframe (gcf));
    end
  close (w)
Is=Is -0.5;
fid=fopen('t0','w');
fprintf(fid,'%f\n',Is);
fclose(fid);
I=I-0.5;
fid=fopen('t1','w');
fprintf(fid,'%f\n',I);
fclose(fid);
```

They are two sequences with 16 frames of size 16×16 each. The selected frames of the sequences are shown in Fig. 5.6. The first row shows the frames of the first sequence with the smooth edge, and the second row shows the second sequence with the sharp edge. A bright area that is a single pixel in the smooth sequence can be a square of any size $d \times d$. In subsequent frames, the entire object with

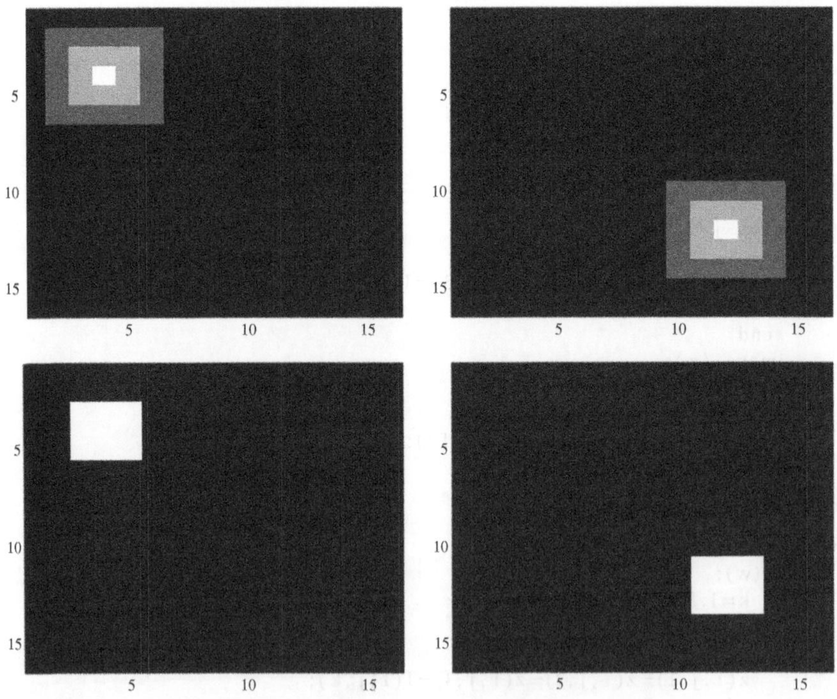

Figure 5.6 4th (first column) and 12th (second column) frames of two artificial videos for smooth and sharp filters.

a bright area changes its position along the diagonal. The function *videoS* produces two files: t0 corresponding to the sequence with a sharp edge, and t1 with a smooth edge. These are taken by the **optimizer** program, which then synthesizes the smooth filter. The sharp filter will be designed after renaming these files: the file t0 should be named t1, and the file t1 should be named t0.

The structure of the 3-D prototype gC circuit is described in the file named **2+2+2C3D3.vhdla** as:

```
library VLSI;
use VLSI.SI.all;
entity example is port (
        terminal output Vout : electrical;
        terminal input Vin : electrical;
        terminal ground gnd : electrical);
end entity example;
-- some comment line
architecture example_arch of example is
variable v1,v2,v3: real;
begin
-- capacitors
C11 : GC_CAP generic map(dim=>1) port map (Ni=>v1, Nj=>v2);
C12 : GC_CAP generic map(dim=>2) port map (Ni=>v1, Nj=>v2);
C13 : GC_CAP generic map(dim=>3) port map (Ni=>v1, Nj=>v2);
C21 : GC_CAP generic map(dim=>1) port map (Ni=>v1, Nj=>v3);
C22 : GC_CAP generic map(dim=>2) port map (Ni=>v1, Nj=>v3);
C23 : GC_CAP generic map(dim=>3) port map (Ni=>v1, Nj=>v3);
C31 : GC_CAP generic map(dim=>1) port map (Ni=>v2, Nj=>v3);
C32 : GC_CAP generic map(dim=>2) port map (Ni=>v2, Nj=>v3);
C33 : GC_CAP generic map(dim=>3) port map (Ni=>v2, Nj=>v3);
-- conductances with I/O signals
G1 : GC_CON port map (Ni=>v1, Nj=>gnd, Vo=>Vout);
G2 : GC_CON port map (Ni=>Vin, Nj=>v3);
-- gyrators
ig1 : GC_GYR port map( Ni=>v2, Nj=>v1);
ig2 : GC_GYR port map( Ni=>v3, Nj=>v2);
ig7 : GC_GYR port map( Ni=>v3, Nj=>v1);

end architecture;
```

According to the *VHDLA* language syntax, terminals are declared in the file header as V_{in}, V_{out}, and *gnd*. In the body of the architecture, there are declared, as the variables, the node voltages V_1, V_2, and V_3 of the circuit and then all circuit elements are listed. The order of the element's listing is arbitrary. The name of the capacitance begins with the letter **C**, followed by the dimension assigned to the capacitance and the nodes between which it is connected. In each dimension, there are three capacitances forming a loop, which means that the filter is of the second order in each of these dimensions. Similarly, the gyrators are connected with their inputs and outputs in a loop, and each of the remaining nodes is connected to the ground. All of the ideal gyrators are described by sequences of characters beginning with the letters **ig**.

In the **gCstudio** environment, the file **2+2+2C3D3.sn** containing the structural numbers (SNs) of the **2+2+2C3D3.vhdl** circuit is generated.

According to the options described in Section 1.1.2, running **optimizer** as follows:

```
for ((i=0; i<=9; i++));
do
./optimizer -c 1 -x -n -t 1 -a t -i 500 -d 16 16 16 2+2+2C3D3.sn
                                                   initialsImag3 > "f$i"
grep Goal "f$i"
done
```

allows to find the values of the parameters of the prototype 3D gC circuit with the structure **2+2+2C3D3.vhdl**. The solutions obtained in the next steps are placed in the files fi, $i = 0, \dots, 9$ using the shift operator >. The **grep** command allows to monitor the progress in the calculations on the monitor screen. The designer decides on the choice of solution, guided mainly by the value of the objective function saved together with the values of the circuit parameters in the files. In Fig. 5.7, the IIR 3D filter synthesis procedure, the role of the environments **gCstudio** and **NANOstudio** in this procedure, and the tools that composed it are shown.

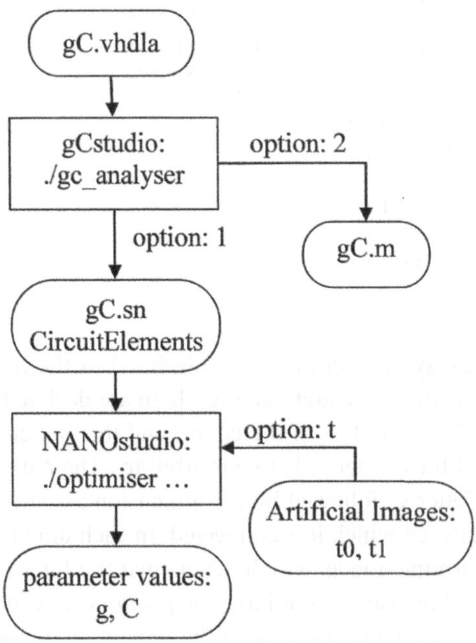

Figure 5.7 Scheme diagram of IIR 3D filter synthesis.

The sharp filter was realized for the series of input samples corresponding to the upper video and the series of output samples corresponding to the lower video from Fig. 5.6, taking a bright area of size 4 × 4 and obtaining an objective function value of 59.5. For the smooth filter, taking the series of input and output samples inversely and the bright region of 2 × 2, the objective function value 29.7 was achieved. The files of the samples, labeled *t*0 and *t*1, respectively, are generated in Octave using the function *videoS*.

The smooth filter was obtained as the counterpart network of gC circuit **2+2+2C3D3.vhdla** for the following parameter values:

```
0.085555901232393
0.000012406860124
0.000050912180339
0.000525416637238
0.000000927452853
0.000012113860834
0.000755193733603
0.000118196676834
0.000155385864249
0.365320299470861
0.999885886245721
0.463385868265641
0.206367200000000
0.849513741749433
```

The first nine values are parameters of the capacitances C_{11}, C_{12}, C_{13}, C_{21}, C_{22}, C_{23}, C_{31}, C_{32}, C_{33}, the next three are transconductances of the gyrators ig_1, ig_2, ig_7, and the last two values are parameters of conductances G_1 and G_2.

The coefficients of the polynomials in the denominator and in the numerator of the gC prototype circuit transfer function

$$H^a(s_1, s_2, s_3) = \frac{P(s_1, s_2, s_3)}{Q(s_1, s_2, s_3)}, \tag{5.19}$$

are presented in the files

3			
2	2	2	
0	0	0	0.31969526
0	0	1	0.00051354
0	0	2	0.00000004
0	1	0	0.00026201
0	1	1	0.00000004
0	2	0	0.00000001
1	0	0	0.39810154
1	0	1	0.00006118
1	1	0	0.00004338
2	0	0	0.00046442

and

3			
2	2	2	
0	0	0	0.31030920
0	0	1	0.00034536
0	0	2	0.00000004
0	1	0	0.00019727
0	1	1	0.00000003
0	2	0	0.00000001
1	0	0	0.21376717
1	0	1	0.00004923
1	1	0	0.00003490
2	0	0	0.00037365

respectively.

The member function *writePoly* of the class **MDp** is used to generate the above files. In addition to the transfer function coefficients, these files contain the circuit dimension (the number 3 in the first line) and the orders (the numbers 2 in the second line) of the numerator and the denominator polynomials in each dimension. In the following lines, the first three numbers denote the power exponents of the variables s_1, s_2, s_3, while the last number is the value of the coefficient. If any of the coefficients is zero, the corresponding line can be omitted. The coefficients are scaled according to the relation (2.20). In direct form, these polynomials are then:

$$\begin{aligned} Q(s_1, s_2, s_3) &= 0.31969526 + 0.00051354 s_3 + \cdots \\ &\quad + 0.00004338 s_1 s_2 + 0.00046442 s_1^2 \\ P(s_1, s_2, s_3) &= 0.31030920 + 0.00034536 s_3 + \cdots \\ &\quad + 0.00003490 s_1 s_2 + 0.00037365 s_1^2. \end{aligned} \tag{5.20}$$

The coefficients, obtained after tensor transformation (2.35), of the polynomials in the denominator and in the numerator of the transfer function

$$H(z_1, z_2, z_3) = \frac{A(z_1, z_2, z_3)}{B(z_1, z_2, z_3)}, \tag{5.21}$$

of the designed IIR 3D video filter, are presented in the files

3			
2	2	2	
0	0	0	0.71914142
0	0	1	1.43713331
0	0	2	0.71799189
0	1	0	1.43767202
0	1	1	2.87304497
0	1	2	1.43537307
0	2	0	0.71853060
0	2	1	1.43591166
0	2	2	0.71738118
1	0	0	0.64001292
1	0	1	1.27797139

1	0	2	0.63795871
1	1	0	1.27897763
1	1	1	2.55384660
1	1	2	1.27486944
1	2	0	0.63896471
1	2	1	1.27587521
1	2	2	0.63691074
2	0	0	−0.07727081
2	0	1	−0.15544647
2	0	2	−0.07817554
2	1	0	−0.15497896
2	1	1	−0.31176752
2	1	2	−0.15678832
2	2	0	−0.07770815
2	2	1	−0.15632105
2	2	2	−0.07861277

and

3			
2	2	2	
0	0	0	0.52507687
0	0	1	1.04936445
0	0	2	0.52428764
0	1	0	1.04968929
0	1	1	2.09780002
0	1	2	1.04811096
0	2	0	0.52461243
0	2	1	1.04843557
0	2	2	0.52382332
1	0	0	0.62095642
1	0	1	1.24053109
1	0	2	0.61957490
1	1	0	1.24112368
1	1	1	2.47948408
1	1	2	1.23836088
1	2	0	0.62016726
1	2	1	1.23895299
1	2	2	0.61878598
2	0	0	0.09737423
2	0	1	0.19415604
2	0	2	0.09678192
2	1	0	0.19442368
2	1	1	0.38766262
2	1	2	0.19323918
2	2	0	0.09704944
2	2	1	0.19350658
2	2	2	0.09645726

respectively. In direct form, these polynomials are then:

$$B(z_1, z_2, z_3) = 0.71914142 + 1.43713331z_3^{-1} + \cdots$$

$$
\begin{aligned}
&- 0.15632105z_1^{-2}z_2^{-2}z_3^{-1} - 0.07861277z_1^{-2}z_2^{-2}z_3^{-2}\\
A(z_1, z_2, z_3) = {}& 0.52507687 + 1.04936445z_3^{-1} + \cdots\\
&+ 0.19350658z_1^{-2}z_2^{-2}z_3^{-1} + 0.09645726z_1^{-2}z_2^{-2}z_3^{-2}.
\end{aligned} \tag{5.22}
$$

The sharp filter was obtained as the counterpart network of the same as for smooth filter gC circuit **2+2+2C3D3.vhdla** structure with the following parameter values:

```
0.657554372630066
0.000083621584633
0.000184990737633
0.046426190520210
0.000003552825649
0.000020753452003
0.000798860745148
0.000076143543499
0.000006917510646
-0.354473245077235
-0.473453896602535
0.637495288282158
0.064722756916686
0.379927702035949
```

The files with the coefficients of the transfer function polynomials have a form analogous to the files obtained in the case of the smooth filter.

Filtering of each 3D block is done in two directions: in the direction of moving frames, and then in the opposite direction. The division of the filtered sequence into 3D blocks, or their integration, is done using the own functions presented above and elaborated in Octave. Since the filter is second order in each dimension, all 3D blocks that make up the sequence contain a 2-pixel-thick border around each frame, which is discarded when the blocks are integrated, and the entire sequence is made up of 3D blocks containing $12 \times 12 \times 16$ pixels.

The designed filters were tested on various videos, coastguard.avi, gbicycle.avi, gflower.avi, used in many works to illustrate sequence processing, e.g. http://www.cs.tut.fi/~foi/GCF-BM3D/ (Maggioni et al., 2014). From the coastguard.avi, there were captured the blocks t*Iij*, $i = 1, \dots, 10, j = 1, \dots, 10$, using the function *videoCap*. These blocks have been filtered

```
for i in $(seq 1 1 10)
do
        for j in $(seq 1 1 10)
        do
        cp "tI$i $j" tI
    ./Fi2 2+2+2C3D3.sn fOut14p4Ha 16 16 16
        cp tOut "tOut$i $j"
        rm tI
        rm tOut
        done
done
```

where **2+2+2C3D3.sn** is obtained in the symbolic analysis of the prototype filter **2+2+2C3D3.vhdla**, **fOut14p4Ha** contains the values of the parameters of the sharp filter obtained using **optimizer** and the last three arguments of **Fi2** are the block sizes. The results of filtering are blocks tOut*ij*, $i = 1, \dots, Nr$, $j = 1, \dots, Nc$, integrated into a new sequence using the function *videoInt*. The smooth filter is similarly tested in this experiment.

In the figures, there are presented eight subsequent frames of the above-obtained videos, which are a part of the coastguard.avi file. Figure 5.8 presents the original video, Figs. 5.9 and 5.10 present the videos after sharp and smooth filtering, respectively.

5.3 Hardware Implementation of 3D Filters

The development of video processing algorithms is closely related to the computing power growing of digital components such as processors, memories, FPGA matrices, etc. FPGAs play a special role as hardware accelerators (HajiRassouliha et al., 2018). The 3D IIR filters presented in Sections 4.1, 4.2, 4.4, 4.6 and 5.2 were implemented in software, with the use of appropriate tools created in the C++ language. As it was discussed, all the blocks into which the sequence is divided, can be processed in parallel. It is the fundamental property that will be used in the hardware implementation presented in this section.

To investigate the possibility of using the presented algorithm in real-time systems, its implementation in the hardware description language Verilog has been prepared. Verilog was chosen because its C++-like syntax made hardware description and verification much easier using the **MDp** class functions presented in Section 2.4.

As part of the implementation, a single filter for video block processing was initially synthesized using the ISE WebPack 14.7 tool. This system was implemented on the Xilinx XC7A200T FPGA matrix, chosen due to the massive amount of available resources and maximum operating frequency. The synthesis succeeded, giving a digital circuit operating at a maximum frequency of 90 MHz using the resources given in Table 5.4.

The symbol of the implemented filter is shown in Fig. 5.11, where the input signals are:

clk – system clock,
r – signal allowing to read the results data *y*,
w – signal allowing to write the input data *x*,
en – signal enabling the system operation,
p – signal starting the 3D block processing,

Figure 5.8 Original frames, from 8th to 15th, of coastguard.avi video.

Figure 5.9 Frames from 8th to 15th of coastguard.avi video, filtered with the sharp filter.

Figure 5.10 Frames from 8th to 15th of coastguard.avi video, filtered with the smooth filter.

Table 5.4 Results of synthesis of 1 and 20 filters on XC7A200T.

	Used/available	
	1 Block	20 Blocks
No. of Slice Registers	603/269200(0%)	2060/269200(4%)
No. of Slice LUTs	1409/134600(1%)	28509/134600(21%)
No. of fully used LUT-FF pairs	404/1608(25%)	8115/32454(25%)
No. of bonded IOBs	71/285(24%)	95/285(33%)
No. of Block RAM/FIFO	9/365(2%)	180/365(49%)
No. of BUFG/BUFGCTRLs	1/32(3%)	1/32(3%)
No. of DSP48E 1s	11/740(1%)	220/740(29%)

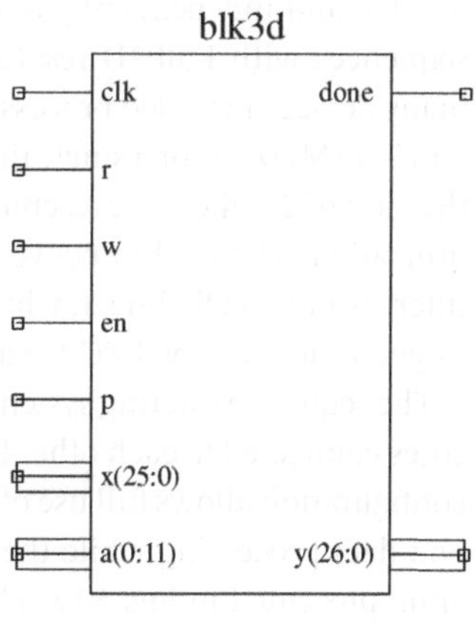

Figure 5.11 Symbol of the filter, implemented on the FPGA, for video processing with marked input (left) and output (right) signals.

x – signal transmitting the input data to the system, and
a – the address of the read/written data from/to the system,

and output signals:

y – a signal transmitting the result data from the system,
done – a signal indicating the end of the block processing.

The filter has its internal memory based on Block RAM and LUT, which speeds up its operation and relieves the data bus. The values of transfer function coefficients of the 3D filter are stored in the built-in memories, along with the values of individual pixels and the processing results.

Video filtering, using the presented system, consists of the following stages:

1. Triggering the circuit – the control block sets the *en* signal high.
2. Write data to the filter – the control unit sets the *w* signal to high, and the *r* and *p* signals to low, *a* to 0 and subsequently sends 4096 pixels contained in the 3D block ($16 \times 16 \times 16$) of the video to the input *x*, increasing the value of the *a* signal by 1.
3. Data processing by the filter – the control unit sets the *p* signal high, the *r* and *w* signals low, and waits for the *done* signal to be high, indicating the end of processing.
4. Reading the results from the filter – the control block sets the *r* signal to high, the *w* and *p* signals to low, and the *a* signal to 0, incrementing the value of the signal *a* reads 4096 values from the *y* signal.

On the selected FPGA there were implemented 20 designed filters operating in parallel and independently of each other. Such a filter system allows processing sequences with FullHD resolution at a speed of 30 fps, which is a standard for many devices for video processing. The maximum operating frequency of the system is 90 MHz, as for a single filter. The usage of the resources of the XC7A200T for the case of 20 filters is presented in Table 5.4. It is worth noting that the resource utilization relationship between a single filter and this system composed of 20 filters is practically linear, which allows to predict the occupancy rate of the prototype system on the FPGA platform.

The sequence filtering system, shown in Fig. 5.12, was built on the basis of single cores connected to each other by a controller supervising memory and cores. This configuration allows full use of the data bus in parallel processing. Memory access and data processing while the system is running are shown in a simplified diagram presented in Fig. 5.13. The division of the system into single cores makes it

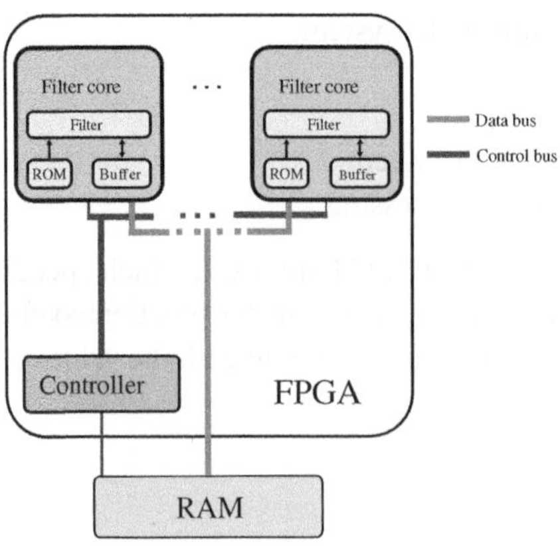

Figure 5.12 The architecture of the digital circuit implementing parallel image filtering in the FPGA.

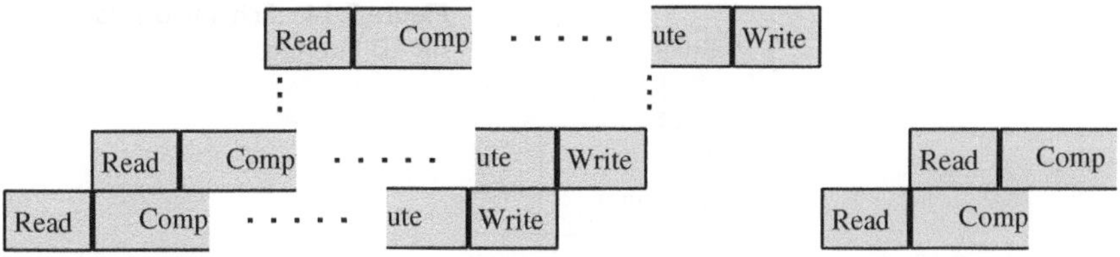

Figure 5.13 Timeline of tasks performed by individual cores in the FPGA.

easier to manage access to RAM by the controller and to optimize the use of available resources of the reprogrammable matrix. Memory is therefore used practically during the entire filtration process, without collisions in memory access by individual blocks, as illustrated in Fig. 5.13. The implementation of the 3D block uses pipelined and parallel processing, described behaviorally. Because a new pixel is generated in each cycle, it is possible to estimate the filtration time for any amount of data. Three states can be distinguished in the operation of the core:

1. Read – the state gives the core access to the RAM in a read mode. At this time, the 3D block of the sequence is loaded from RAM into the core's internal buffer. Thanks to this, each core is isolated from memory during calculations.
2. Compute – in the state the 3D block read in the previous step is filtered. Processing is performed in internal memory independently of RAM and other cores.
3. Write – the state rewrites the filter results from the internal core buffer to RAM. At the same time, the core terminates its operation and waits for it to return to the Read state.

The Read and Write states are completely supervised by the Controller, so memory is always in use as shown in Fig. 5.13, and memory is not accessed by more than one core at a time. Tasks are assigned to each core in such a way as not to block the data bus. This guarantees optimal access to memory and parallelization of calculations. This means that every time some core starts its individual calculation, some other core reads data from memory or writes data to memory. Pipelining is thus achieved.

FPGA implements the operation of the **Fi2** filter in hardware in the **NANOstudio** environment as it is shown in Fig. 5.14. Depending on the computing power of the digital system used, an appropriate part of the total number of $Nr \times Nc$ blocks can be processed in parallel. As shown in Table 5.4, it is 20 such filters were implemented on XC7A200T. Only the filter operation in Section 5.2 was performed individually for each block.

Let us note that the hardware implementation of the algorithm, in relation to the presented in Section 5.2, required to introduce a change related to the representation of floating-point numbers to Q3.24 fixed-point numbers. This change obviously affects the accuracy of the result. For the single $16 \times 16 \times 16$ block, the Mean

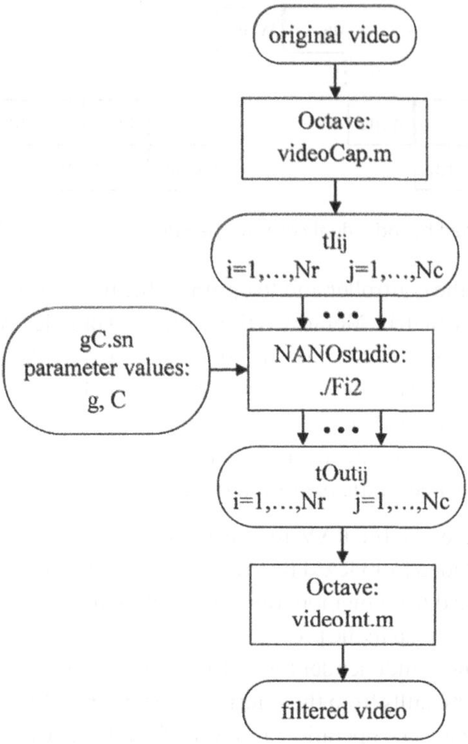

Figure 5.14 Scheme diagram of IIR 3D filtering.

Squared Error (MSE) was 2.33229670410156E-08, comparing the results from the FPGA matrix with the results generated by C++ tools in Section 5.2. It is worth mentioning that the result is also influenced by replacing the operation of division with multiplication by the reciprocal of the divisor. These treatments slightly worsened the MSE, but improved the operating frequency of the system as much as six times, hence, the achieved maximum operating frequency is 90 MHz.

5.4 3D Filters in Video Coding

Filtering is used at many stages of video encoding and decoding. Filters play a significant role in the process of reconstruction of removed pixels. Another example of the use of filters is finding subpixels and intermediate pixels at the stage of calculating the MV. One-dimensional FIR filters are used for that purpose (Han et al., 2021). This section, will show the usefulness of the 3D IIR filters discussed in this chapter for finding arrays of intermediate pixels and subpixels.

In the example presented, the ideal video is reduced by removing, in each frame, pixels in even rows and even columns. The initial video obtained in this way will then be extended by intermediate pixels and subpixels to the size of an ideal video, according to the following algorithm:

1. a sequence of intermediate pixels is created by sharpening the initial video,
2. the initial video is extended with columns taken from the sequence of intermediate pixels,
3. a sequence consisting of subpixels is created by sharpening the column-extended video,
4. the initial video augmented in the columns in step 2 gives the final video after increasing again with the rows composed of subpixels obtained in step 3.

The operations on the sequences in the above algorithm are performed using the following proprietary functions developed in the Octave environment:

Com2D – removes even rows and columns in the ideal sequence,
DeComCol – creates the extended sequence in step 2 of the algorithm,
DeComRow – creates the extended sequence in step 4 of the algorithm.

As an ideal video we'll take gbicycle.avi with dimensions of 576 rows, 720 columns, and 30 frames ($576 \times 720 \times 30$), using the **Fi2** sharp filter from Section 5.2 to calculate intermediate pixels and subpixels. According to the general description of the algorithm above, the next steps in the sequence processing are as follows:

1. *Com2D* – function removes the even columns and rows in every frame of gbicycle.avi ($576 \times 720 \times 30$) presented in Fig. 5.15a, producing the initial video comBic2.avi ($288 \times 360 \times 30$) presented in Fig. 5.15b,
2. *videoCap* – function captures, from comBic2.avi, tIij blocks, where $i = 1, \dots, 23$, $j = 1, \dots, 29$, and number 23×29 of blocks is calculated from the formula (5.18): $\lfloor (288 - 2 \cdot 2)/12 \rfloor = 23$, $\lfloor (360 - 2 \cdot 2)/12 \rfloor = 29$,
3. filtering tIij – blocks with **Fi2** produces the intermediate pixel blocks tOutij, for i, j as in p. 2,
4. resulting tOutij – blocks are integrated, similarly to tIij, with the function *videoInt* into a sequence of intermediate pixels inCom1H4.avi ($276 \times 348 \times 16$), for i, j as in p. 2,
5. *DeComCol* – function creates the column-extended sequence DeComH4c.avi ($276 \times 696 \times 16$), taking successive columns from inCom1.avi as odd columns and from inCom1H4.avi as even columns,
6. *videoCap* – captures, from DeComH4c.avi, blocks tIij, $i = 1, \dots, 22, j = 1, \dots, 57$, integrated by *videoInt* into the video inCom2.avi ($264 \times 684 \times 16$), where i and j are calculated as $\lfloor (276 - 2 \cdot 2)/12 \rfloor = 22$, $\lfloor (696 - 2 \cdot 2)/12 \rfloor = 57$,

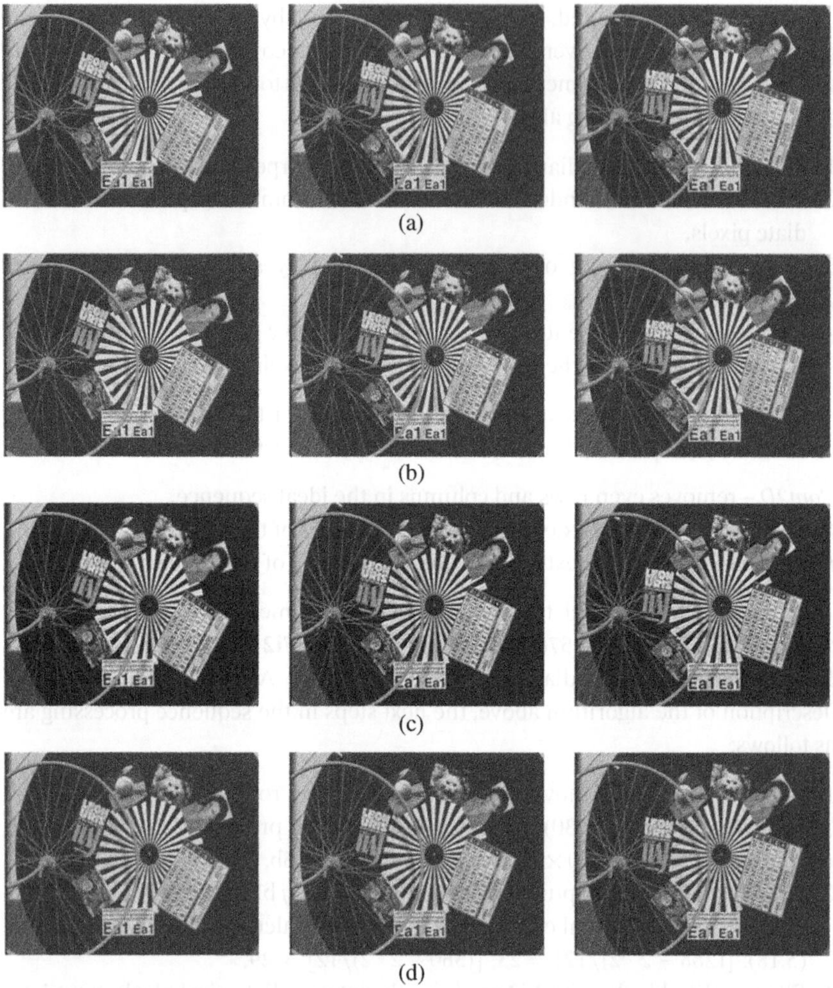

Figure 5.15 Frame numbers 2, 8, and 14 of the gbicycle.avi video in ideal (a), initial (b) and in the final versions after extension by intermediate pixels and subpixels using sharp (c) and averaging (d) filters.

7. tI*ij* – blocks are processed by **Fi2** filter, for *i*, *j* as in p. 6,
8. resulting blocks of subpixels tOut*ij* are integrated by *videoInt* into the video inCom2H4.avi (264 × 684 × 16), for *i*, *j* as in p. 6,
9. the *DeComRow* function creates the row-extended video DeComH4r.avi (528 × 684 × 16), taking the odd rows from inCom2.avi and the even rows from inCom2H4.avi.

Selected frames appearing in the above process steps are shown in Fig. 5.15. The frames: 2nd, 8th, and 14th, taken from the ideal video gbicycle.avi, are presented in Fig. 5.15a. These frames are slightly reduced due to block capture according to equation (5.18). Figure 5.15b shows the same frames in the initial video, and Fig. 5.15c in the final video, after expanding with intermediate pixels and subpixels obtained after the sharp filtering. On the other hand, Fig. 5.15d shows for comparison the final sequence obtained using the bilinear averaging filter, i.e. the first-order FIR filter with coefficients [1/2, 1/2] implemented in the *DeComE2* function. The selected patches, from the different versions of the 14th frame in the last column in Fig. 5.15, are shown in Fig. 5.16. In subjective evaluation, the sharp 3D filter improves the quality of the final sequence, while the sequence obtained using the averaging filter shows its blurring. The use of higher-order FIR filters available, for example, in the AV1 specification (de Rivaz and Haughton, 2022),

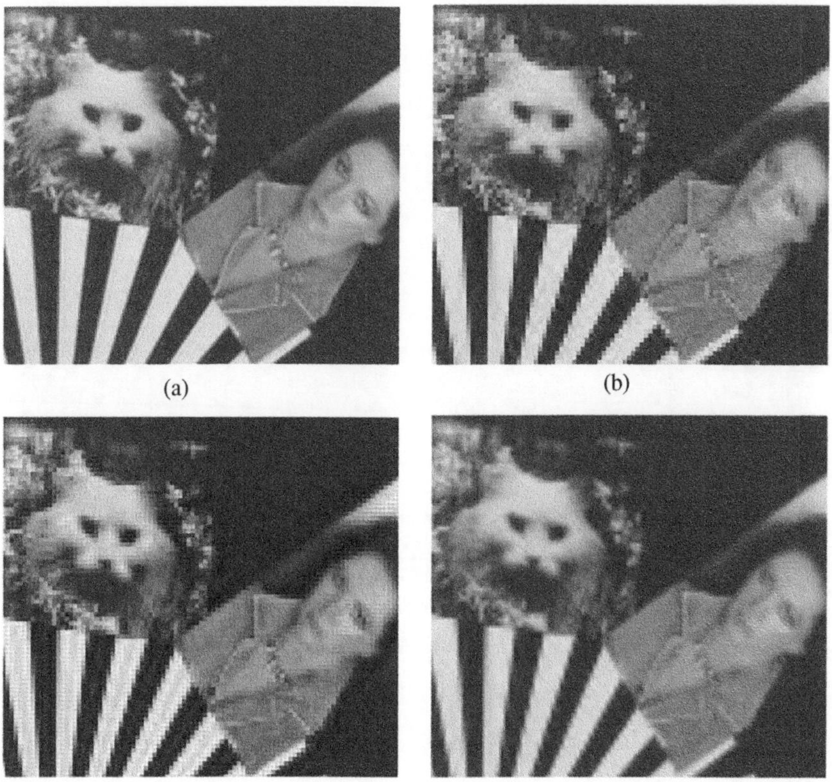

(a) (b) (c) (d)

Figure 5.16 Patches of the 14th frame (last column in Fig. 5.15) of the gbicycle.avi in ideal (a), initial (b) and final versions obtained after sharp (c) and averaging (d) filtering.

not improves the quality of the final sequence. The objective evaluation, using the Structural SIMilarity (SSIM) index, gives comparable results for both presented cases. For the averaging filter, the SSIM is 0.932, while for the sharp filter it is 0.928. The indices were calculated using the *SSIM* function, which implements for videos the formulae given in Wang et al. (2004). In conclusion, we state that intermediate pixels and subpixels are calculated with greater precision using the IIR sharp filter than using the FIR filters.

Bibliography

O. T. Altinoz and A. E. Yilmaz. Parallelization of Hooke-Jeeves pattern recognition algorithm by using CUDA for GPGPU. *Signal Processing and Communications Applications Conference (SIU)*, pages 1793–1796, 2015.

J. Bao and Z. Kang. A family of binomial polynomials and its applications. *Journal of Electronic Science and Technology of China*, 4(3):249–252, 2006.

St. Bellert. Topological analysis and synthesis of linear systems. *Journal of the Franklin Institute*, 274:377–443, 1962.

I. Bender, A. Borges, L. Agostini, B. Zatt, G. Correa, and M. Porto. Complexity and compression efficiency analysis of *libaom* AV1 video codec. *Journal of Real-Time Image Processing*, 20(3):50, 2023. doi: https://doi.org/10.1007/s11554-023-01308-5.

V. Bhaskaran and K. Konstantinides. *Image and Video Compression Standards.* Kluwer Academic Publishers, Boston, MA, 1995.

C. E. Braun, L. D. Chiwiacowsky, and A. T. Gómez. Variations of ant colony optimization for the solution of the structural damage identification problem. *Procedia Computer Science*, 51:875–884, 2015.

P. Bu, H. Zhao, Y. Jin, and Y. Ma. Linear recursive non-local edge-aware filter. *IEEE Transactions on Circuits and Systems for Video Technology*, 31(5):1751–1763, 2021.

T. Chen, H. R. Wu, and B. Qiu. Adaptive postfiltering of transform coefficients for the reduction of blocking artifacts. *IEEE Transactions on Circuits and Systems for Video Technology*, 11(5):594–602, 2001.

R. M. Foster. A reactance theorem. *Bell System Technical Journal*, 3(2):259–267, April 1924.

L. Gazsi. Explicit formulas for lattice wave digital filters. *IEEE Transactions on Circuits and Systems*, 32(1):68–88, 1985.

B. Gilbert. A precise four-quadrant multiplier with subnanosecond response. *IEEE Journal of Solid-State Circuits*, 3(4):365–373, 1968.

A. HajiRassouliha, A. J. Taberner, M. P. Nash, and P. M. F. Nielsen. Suitability of recent hardware accelerators (DSPs, FPGAs, and GPUs) for computer vision and

Methods and Tools for Computer-aided Design of Multidimensional Filters in Signal Processing,
First Edition. Andrzej Handkiewicz.

image processing algorithms. *Signal Processing: Image Communication*, 68:101–119, 2018.

J. Han, B. Li, D. Mukherjee, C.-H. Chuiang, A. Grange, C. Chen, H. Su, S. Parker, S. Deng, U. Joshi, Y. Chen, Y. Wang, P. Wilkins, Y. Xu, and J. Bankoski. A technical overview of AV1. *Proceedings of the IEEE*, 109(9):1435–1462, 2021.

A. Handkiewicz. Two-dimensional SC filter design using a gyrator-capacitor prototype. *International Journal of Circuit Theory and Applications*, 16:101–105, 1988.

A. Handkiewicz. Two-dimensional switched capacitor filter design system for real-time image processing. *IEEE Transactions on Circuits and Systems for Video Technology*, 1(3):241–246, 1991.

A. Handkiewicz. *Mixed-Signal Systems: A Guide to CMOS Circuit Design*. Wiley-Interscience, IEEE Press, New York, 2002.

A. Handkiewicz and M. Naumowicz. NANO-studio, the design environment of filter banks implemented in standard CMOS technology. *Analog Integrated Circuits and Signal Processing*, 109(2):323–333, 2021.

A. Handkiewicz, P. Katarzynski, S. Szczesny, J. Wencel, and P. Sniatala. Analog filter pair design on the basis of a gyrator-capacitor prototype circuit. *International Journal of Circuit Theory and Applications*, 40:539–550, 2012.

A. Handkiewicz, P. Katarzynski, S. Szczesny, M. Naumowicz, M. Melosik, P. Sniatala, and M. Kropidlowski. Design automation of a lossless multiport network and its application to image filtering. *Expert Systems with Applications*, 41(5):2211–2221, 2014.

A. Handkiewicz, S. Szczesny, M. Naumowicz, P. Katarzynski, M. Melosik, P. Sniatala, and M. Kropidlowski. SI-Studio, a layout generator of current mode circuits. *Expert Systems with Applications*, 42(6):3205–3218, 2015.

P. G. Howard and J. S. Vitter. Arithmetic coding for data compression. *Proceedings of the IEEE*, 82(6):857–865, 1994.

C.-Y. Huang, G. Dewey, E. Mannebach, A. Phan, P. Morrow, W. Rachmady, I. C. Tung, N. Thomas, U. Alaan, R. Paul, and N. Kabir. 3-D self-aligned stacked NMOS-on-PMOS nanoribbon transistors for continued Moore's law scaling. *IEEE International Electron Devices Meeting (IEDM)*, pages 20.6.1–20.6.4, 2020.

J. B. Hughes, K. W. Moulding, J. Richardson, J. Bennett, W. Redman-White, M. Bracey, and R. S. Soin. Automated design of switched-current filters. *IEEE Journal of Solid-State Circuits*, 31(7):898–907, 1996.

A. K. Jain. *Fundamentals of Digital Image Processing*. Prentice Hall, Inc., Hoboken, New Jersey, 1989.

D. Johnson. Big blue gets small IBM's 2-nanometer chip is a world's first. *IEEE Spectrum*, 8:7, 2021.

P. Läwenborg, H. Johansson, and L. Wanhammar. Two-channel digital and hybrid analog/digital multirate filter banks with very low-complexity analysis or synthesis filters. *IEEE Transactions on Circuits and Systems-II: Analog and Digital Signal Processing*, 50(7):355–367, 2003.

R. G. Lyons. *Understanding Digital Signal Processing*. Prentice Hall PTR, Hoboken, New Jersey, 2001.

J. P. Magalhaes, J. M. N. Vieira, R. Gómez-Garcia, and N. B. Carvalho. Bio-inspired hybrid filter bank for software-defined radio receivers. *IEEE Transactions on Microwave Theory and Techniques*, 61(4):1455–1466, 2013.

M. Maggioni, G. Boracchi, A. Foi, and K. Egiazarian. Video denoising, deblocking, and enhancement through separable 4-D nonlocal spatiotemporal transforms. *IEEE Transactions on Image Processing*, 21(9):3952–3966, 2012.

M. Maggioni, E. Sanchez-Monge, and A. Foi. Joint removal of random and fixed-pattern noise through spatiotemporal video filtering. *IEEE Transactions on Image Processing*, 23(10):4282–4296, 2014.

F. Maloberti. *Analog Design for CMOS VLSI Systems*. Kluwer Academic Publishers, Boston, Massachusetts, 2001.

M. Naumowicz, M. Melosik, P. Katarzynski, and A. Handkiewicz. Automation of CMOS technology migration illustrated by RGB to YCrCb analogue converter. *Opto-Electronics Review*, 21:326–331, 2013.

A. Norkin, G. Bjontegaard, A. Fuldseth, M. Narroschke, M. Ikeda, K. Andersson, M. Zhou, and G. Van der Auwera. HEVC deblocking filter. *IEEE Transactions on Circuits and Systems for Video Technology*, 22(12):1746–1754, 2012.

A. V. Oppenheim, R. W Schafer, and J. R. Buck. *Discrete-Time Signal Processing*. Prentice Hall, Upper Saddle River, New Jersey, 1999.

H. J. Orchard. Inductorless filters. *Electronics Letters*, 2:224–225, 1966.

H. J. Orchard, G. C. Temes, and T. Cataltepe. Sensitivity formulas for terminated lossless two-ports. *IEEE Transactions on Circuits and Systems*, 32(5):459–466, 1985.

L. Qiu, Y. J. Zheng, and L. Siek. Design of frequency-interleaved ADC with mismatch compensation. *Electronics Letters*, 50(9):659–661, 2014.

D. Raychaudhuri and N. B. Mandayam. Frontiers of wireless and mobile communications. *Proceedings of the IEEE*, 100(4):824–840, 2012.

P. de Rivaz and J. Haughton. AV1 bitstream & decoding process specification. Available: https://aomediacodec.github.io/av1-spec/av1-spec.pdf, Online: Accessed: 2022, 2022.

R. Rudnicki, M. Kropidłowski, and A. Handkiewicz. Low power switched-current circuits with low sensitivity to the rise/fall time of the clock. *International Journal of Circuit Theory and Applications*, 38:471–486, 2010.

E. Sanchez-Sinencio and A. G. Andreou. *Low-Voltage/Low-Power Integrated Circuits and Systems: Low-Voltage Mixed-Signal Circuits*. Wiley-IEEE Press, 1999.

T. Saramaki. On the design of digital filters as a sum of two all-pass filters. *IEEE Transactions on Circuits and Systems*, 32(11):1191–1193, 1985.

M. F. Simón Gálvez, S. J. Elliott, and J. Cheer. Time domain optimization of filters used in a loud-speaker array for personal audio. *IEEE/ACM Transactions on Audio, Speech and Language Processing*, 23(11):1869–1878, 2015.

P. Sniatala, M. Naumowicz, A. Handkiewicz, S. Szczesny, J. de Melo, N. Paulino, and J. Goes. Current mode sigma-delta modulator designed with the help of transistor's size optimization tool. *Bulletin of the Polish Academy of Sciences Technical Sciences*, 63:919–922, 2015.

S. Szczesny. Current-mode FPAA with CMRR elimination and low sensitivity to mismatch. *Circuits, Systems, and Signal Processing*, 36(7):2672–2696, 2017.

S. Szczesny, M. Naumowicz, and A. Handkiewicz. SI-Studio: environment for SI circuits design automation. *Bulletin of the Polish Academy of Sciences Technical Sciences*, 60:757–762, 2012.

B. D. H. Tellegen. The gyrator, a new electric network element. *Philips Research Reports*, 3(4):81–101, 1948.

G. C. Temes and J. LaPatra. *Introduction to Circuit Synthesis and Design*. McGraw-Hill, Inc., New York, 1977.

W.-C. Tu and S.-Y. Chien Two-way recursive filtering. *IEEE Transactions on Circuits and Systems for Video Technology*, 31(11):1435–4268, 2021.

Z. Wang, A. C. Bovik, H. R. Sheikh, and E. P. Simoncelli. Image quality assessment: From error visibility to structural similarity. *IEEE Transactions on Image Processing*, 13(4):600–612, 2004.

A. N. Willson and H. J. Orchard. Insights into digital filters made as the sum of two allpass functions. *IEEE Transactions on Circuits and Systems I*, 42(3):129–137, 1995.

E. J. Wyers, M. A. Morton, T. C. L. G. Sollner, C. T. Kelley, and P. D. Franzon. A generally applicable calibration algorithm for digitally reconfigurable self-healing RFICs. *Transactions on Very Large Scale Integration (VLSI) Systems*, 24:1151–1164, 2016.

S. Zhao and S. Chan. Design and multiplierless realization of digital synthesis filters for hybrid-filterbank A/D converters. *IEEE Transactions on Circuits and Systems I: Regular Papers*, 56(10):2221–2233, 2009.

Index

Methods and Tools for Computer-aided Design of Multidimensional Filters in Signal Processing, First Edition. Andrzej Handkiewicz.

Printed and bound by CPI Group (UK) Ltd, Croydon, CR0 4YY

17/09/2026

14964331-0008